SITE PLANNING

Lester Wertheimer, FAIA, with Contributing Editor Laura Serebin, AIA

This publication is designed to provide accurate and authoritative information in regard to the subject matter covered. It is sold with the understanding that the publisher is not engaged in rendering legal, accounting, or other professional service. If legal advice or other expert assistance is required, the services of a competent professional person should be sought.

President: Roy Lipner
Vice President of Product Development and Publishing: Evan M. Butterfield
Editorial Project Manager: Jason Mitchell
Director of Production: Daniel Frey
Production Editor: Caitlin Ostrow
Creative Director: Lucy Jenkins

© 1992, 1998, 2001 by Architectural License Seminars, Inc.
© 2007 by Dearborn Financial Publishing, Inc.®

Published by Kaplan AEC Education
30 South Wacker Drive, Suite 2500
Chicago, IL 60606-7481
(312) 836-4400
www.kaplanAECarchitecture.com

All rights reserved. The text of this publication, or any part thereof, may not be reproduced in any manner whatsoever without permission in writing from the publisher.

Printed in the United States of America.

07 08 09   10 9 8 7 6 5 4 3 2 1

CONTENTS

Introduction   v

LESSON ONE

THE SITE PLANNING DIVISION   1

    Introduction   1
    The Exam Format   1
    The Site Planning Division   2
    Emphasis of the Exam   2
    Grading Criteria   3
    The Question of Failure   3
    What Is Expected   4
    How to Prepare   4
    Practice Exams   5

LESSON TWO

THE NCARB SOFTWARE   7

    Introduction   7
    Vignette Screen Layout   7
    Computer Tools   8

LESSON THREE

TAKING THE EXAM   11

    Introduction   11
    Scheduling the Exam   11
    Final Preparation   11
    Exam Day   12
    The Exam Room   12
    Exam Room Conduct   12
    Begin with the Program   13
    General Strategies   13
    Exam Purpose   14
    The Time Schedule   14
    Time Scheduling Problems   14
    Managing Problems   15
    Working Under Pressure   15
    Examination Advice   15

LESSON FOUR

SITE PLANNING ISSUES: ASSESSING THE SITE   17

    Introduction   18
    Water   19
    Plants   26
    Introduction to Topography   29
    Contours   30
    Grading   36
    Drainage   41
    Design Methods   46
    Site Analysis   53

LESON FIVE

SITE PLANNING ISSUES: DESIGNING THE SITE   59

    Site Design Process   59
    Circulation Systems   60
    Zoning   75
    Flexible Zoning   81

LESSON SIX

SITE DESIGN VIGNETTE   87

    Introduction   87
    Vignette Information   87
    Design Procedure   88
    Analyzing the Site Plan   89
    Important Parking Objectives   90
    Arranging the Elements   91
    Circulation   91
    Final Arrangement   92
    Key Computer Tools   92
    Vignette 1 Site Design   93

LESSON SEVEN

SITE ZONING VIGNETTE   101

Introduction   101
Vignette Information   101
Design Procedure   102
Key Computer Tools   103
Vignette 2 Site Zoning   104

LESSON EIGHT

SITE GRADING VIGNETTE   113

Introduction   113
Vignette Information   113
Design Procedure   114
Key Computer Tools   117
Vignette 5 Site Grading   118

LESSON NINE

PRACTICE VIGNETTES   125

Vignette 1 Buildable Area   127
Vignette 2 Parking Layout   132
Vignette 3 Site Parking   137
Vignette 4 Site Grading   143

LESSON TEN

EXERCISE PROBLEMS   149

Problem 1   150
Problem 2   154
Problem 3   159
Problem 4   164
Problem 5   169
Problem 6   174
Problem 7   179
Problem 8   183
Problem 9   188
Problem 10   193
Problem 11   198
Problem 12   204
Problem 13   210
Problem 14   215
Problem 15   221

Glossary   227
Bibliography   235
Index   237

INTRODUCTION

WELCOME

Thank you for choosing Kaplan AEC Education for your ARE study needs. We offer updates every January to keep abreast of code and exam changes and to address any errors discovered since the previous update was published. We wish you the best of luck in your pursuit of licensure.

ARE OVERVIEW

Since the State of Illinois first pioneered the practice of licensing architects in 1897, architectural licensing has been increasingly adopted as a means to protect the public health, safety, and welfare. Today, all U.S. states and Canadian provinces require licensing for individuals practicing architecture. Licensing requirements vary by jurisdiction; however, the minimum requirements are uniform and in all cases include passing the Architect Registration Exam (ARE). This makes the ARE a required rite of passage for all those entering the profession, and you should be congratulated on undertaking this challenging endeavor.

Developed by the National Council of Architectural Registration Boards (NCARB), the ARE is the only exam by which architecture candidates can become registered in the United States or Canada. The ARE assesses candidates' knowledge, skills, and abilities in nine different areas of professional practice, including a candidate's competency in decision making and overall knowledge of various areas of the profession. The exam also tests competence in fulfilling an architect's responsibilities and for coordinating the activities of others while working with a team of design and construction specialists. In all jurisdictions, candidates must pass the nine divisions of the exam in order to become registered.

The ARE is designed and prepared by architects, making it a practice-based exam. It is generally not a test of academic knowledge, but rather a means to test decision-making ability as it relates to the responsibilities of the architectural profession. For example, the exam does not expect candidates to memorize specific details of the building code, but requires them to understand a model code's general requirements, scope, and purpose, and to know the architect's responsibilities related to that code. As such there is no substitute for a well-rounded internship to help prepare for the ARE.

Exam Format

The ARE consists of nine divisions: three graphic and six multiple-choice. The three graphic divisions of the ARE are composed of a series of vignettes intended to assess knowledge, skills, and abilities in the different facets of architectural practice. Each of the divisions is administered within a fixed time limit. Exam candidates are required to create a graphic solution for each of the vignettes following program and code requirements. The vignette order and time limits for these divisions are outlined in the table below. For detailed information on the multiple-choice divisions, refer to the study guides for those divisions.

Site Planning		
Section 1	1.5 hours	Site Design (15-minute break)
Section 2	1.5 hours	Site Zoning Site Grading
Building Planning		
Section 1	1 hour	Interior Layout (15-minute break)
Section 2	4 hours	Schematic Design
Building Technology		
Section 1	2.5 hours	Building Section Structural Layout Accessibility/Ramp (15-minute break)
Section 2	2.75 hours	Mechanical/Electrical Plan Stair Design Roof Plan

NEW TO THE EXAM

ARE 3.1

In November 2005, NCARB released *ARE Guidelines* Version 3.1, which outlines changes to the exam that are effective as of February 2006. These new guidelines detail several changes for the Site Planning division. Noteworthy points are outlined below.

- The site design and site parking vignettes have been combined into a single "site design" vignette.
- The site zoning and site analysis vignettes have been combined into a single "site zoning" vignette.
- The overall time limits for the Site Planning division remain the same. However, the recommended time to complete the revised vignettes has changed.
- The list of codes and standards candidates should familiarize themselves with for all divisions has been reduced to those of the International Code Council (ICC), the National Fire Protection Association (NFPA), and the National Research Council of Canada.
- All other division statements and content area descriptions remain unchanged.

For details about the multiple-choice divisions, please refer to the study guides for those divisions.

Rolling Clock

A rolling clock went into effect January 1, 2006. Candidates must now pass all nine ARE divisions within a five-year period. Additionally, NCARB has instituted a set of "transitional rules" for candidates already in the process of taking the ARE when the clock went into effect. See the new guidelines or visit the NCARB Web site for more detailed information.

The exam presents vignettes sequentially. Candidates may move between vignettes within a single exam section, allowing them to return to a completed vignette and recheck their work. This also allows candidates some flexibility to slightly adjust the time necessary to complete their vignette solutions within an exam section.

Actual appointment times for taking the exam are somewhat longer than the actual exam time to allow candidates to check in and out of the testing center and to account for the scheduled break. All ARE candidates are encouraged to review NCARB's *ARE Guidelines* for further detail about the exam format, including recommended time allotment for each of the vignettes.

These guidelines are available via free download at the Council's Web site *(www.ncarb.org)*.

Vignette Format

It is important for exam candidates to become familiar not only with exam content, but also question format. Familiarity with the basic question types found in the ARE will reduce confusion, save time, and help you pass. As such, the single most important thing candidates can do to prepare themselves for the graphic divisions is to become fluent in the use of NCARB's graphic software. NCARB has made practice software available that can be downloaded free of charge from their Web site. Candidates should download this software and become thoroughly familiar with its use.

Recommendations on Exam Division Order

NCARB allows candidates to choose the order in which they take the exams, and the choice is an important one. While only you know what works best for you, the following are some general considerations that many have found to be beneficial:

1. The Building Design/Materials & Methods and Pre-Design divisions are perhaps the broadest of all the divisions. Although this can make them among the most intimidating, taking these divisions early in the process will give a candidate a broad base of knowledge and may prove helpful in preparing for subsequent divisions. An alternative to this approach is to take these two divisions last because you will already be familiar with much of their content. This latter approach likely is most beneficial when you take the exam divisions in fairly rapid succession so that details learned while studying for earlier divisions will still be fresh in your mind.

2. The Construction Documents & Services exam covers a broad range of subjects, dealing primarily with the architect's role and responsibilities within the building design and construction team. Because these subjects serve as one of the core foundations of the ARE, it may be advisable to take this division early in the process, as knowledge gained preparing for this exam can help in subsequent divisions.

3. The General Structures and Lateral Forces divisions cover related and overlapping subjects. Take them consecutively, and take General Structures first, because it is broader and addresses fundamental principles necessary for success in Lateral Forces.

4. The three graphic divisions all use an identical software platform and employ similar graphic drawing tools. Because becoming fluent with this software is crucial to passing these exams, take the three graphic divisions sequentially.

5. The Mechanical & Electrical Systems and Building Technology exams cover loosely related material. As such, it is often beneficial to take these two exams consecutively.

6. Take exams that particularly concern you early in the process. NCARB rules prohibit retaking an exam for six months. Therefore, failing an exam early in the process will allow the candidate to use the waiting period to prepare for and take other exams.

EXAM PREPARATION

Overview

There is little argument that preparation is key to passing the ARE. With this in mind, Kaplan has developed complete learning systems for the graphic divisions that include study guides, practice vignettes, a CD-ROM test bank, and flash cards. This study guide offers a condensed course of study and will best prepare you for the exam when utilized along with the other tools in the learning system. The system is designed to provide you with the general background necessary to pass the exam and to provide an indication of specific content areas that demand additional attention.

In addition to the Kaplan learning systems, materials from industry-standard documents may prove useful for the various divisions of the ARE.

Course Method

This manual guides candidates through the Site Planning division of the ARE by familiarizing you with the specifics of the test and reviewing simulated vignette problems. Following each vignette example is a suggested graphic solution, together with an analysis and explanation of how it evolved. Although other solutions are possible, the approach in every case consists of a logical sequence of steps that have proven successful over the years. The principal goal of this study aid is not to be a primer on design, but instead to teach an effective and methodical technique for approaching a difficult and unique examination. Candidates are encouraged to follow the logical process identified in this manual, step by step, in order to better understand the procedure required to successfully solve ARE vignette problems.

In addition to the vignette examples that are typical of the current ARE computerized test, actual NCARB vignettes from previous ARE exams, as well as a number of related exercise problems created by Kaplan AEC Education are included. All of these examples are intended to prepare candidates as completely as possible for the Site Planning graphic exam.

Preparation Basics

The first step in preparation should be a review of the exam specifications and reference materials published by NCARB. These statements are available for each of the nine ARE divisions to serve as a guide for preparing for the exam. Download these statements and familiarize yourself with their content. This will help you focus your attention on the subjects that are the focal point of each exam.

As mentioned, the most important element of preparation for the graphic divisions is to become fluent in the use of NCARB's graphic testing software. The NCARB practice program allows candidates to become familiar with the interface and tools. Developed to assist candidates in preparing to use the ARE's graphic software, this practice program consists of tutorials, instructions, and practice vignettes for each of the eleven vignettes found within the ARE's graphic divisions. Candidates should spend as much time as required to feel comfortable with the use of the software and tools prior to scheduling their exam appointment.

Prior knowledge of CAD or other graphic drawing programs is not necessary to successfully complete the exam. In fact, it is important for candidates familiar with CAD to realize they will experience significant differences between the drawing tools used in the ARE and the commercial CAD software used in practice.

While no two people will have exactly the same ARE experience, the following are recommended best practices to adopt in your studies:

Set aside scheduled study time.
Establish a routine and adopt study strategies that reflect your strengths and mirror your approach in other successful academic pursuits. Most important, set aside a definite amount of study time each week, just as if you were taking a lecture course, and carefully read all of the material.

Utilize the study guide.
After studying the materials in the study guide, practice solving the vignettes found at the conclusion of each lesson. The vignettes are intended to be straightforward and objective. Solutions and explanations can be found within the lessons. Pay special attention to the procedure used to work through each vignette.

Utilize the practice vignettes.
Additional practice vignettes can be found at the end of each lesson, allowing you the opportunity to practice working through different vignettes

and pinpointing areas where you need improvement. Reread and take note of the study guide sections that cover these areas and seek additional information from other sources. If you've purchased the practice vignettes, use them as a final tune-up for the exam.

Practice using the NCARB software.
Work through the practice vignettes contained within the NCARB software. You should work through each of these vignettes repeatedly until you can solve them fluently without any difficulty utilizing the software. As you develop your skills, keep track of how long it takes you to work through a solution for each one and note this for exam day.

Supplementary Study Materials

In addition to the Kaplan learning system, materials from industry-standard sources may prove useful for the various divisions of the ARE. Candidates should consult the list of exam references in the NCARB guidelines for the council's recommendations.

Test-Taking Advice

Preparation for the exam should include a review of successful test-taking procedures—especially for those who have been out of the classroom for some time. Following is advice to aid in your success:

Pace yourself.
Each vignette allows candidates ample time to complete their vignette solutions within the time allotted. You should be able to comfortably read the program requirements before beginning your solution.

Read carefully.
Begin each question by reading it carefully and fully reviewing the instructions and requirements. Make a quick list of the requirements to check your work after completing the vignette.

Budget your time.
Candidates should know before entering the exam room approximately how much time is needed to solve each vignette. We recommend budgeting this time to allow 5–10 minutes to carefully read the instructions and program requirements and 10–15 minutes to review your solution at the end, confirming that all program requirements have been met.

Remember that style doesn't count.
Successful vignette solutions are graded based on their conformance with the program requirements and instructions. Accordingly, candidates should not waste any time attempting to create solutions that are attractive or by adding any features that are not required.

Review your work.
Review your vignette solution and carefully check it against the testing criteria. Make sure that your solution has addressed every vignette requirement.

Take advantage of time flexibility.
Where there are multiple vignettes included in a single exam section, the graphic divisions allow candidates to move between vignettes. If, during your practice, you discover that a particular vignette is consistently causing difficulty, take note of which other vignettes are included in the same exam section and the total permitted time. You may be able to utilize a few extra minutes for a more difficult vignette if you know that another poses significantly less difficulty.

Calculator.
Candidates must bring their own calculator to the testing center. Note that only nonprogrammable, noncommunicating, nonprinting calculators are allowed. Candidates will need only a

basic scientific calculator with trigonometry functions. Calculators capable of storing formulas are not permitted.

Keep an eye on the clock.
Although the ARE does note the elapsed time on the testing screen, there are no alerts or messages to warn you that time is running out. During the graphic exams it is easy for candidates to become absorbed in their solution. You should therefore keep a close watch on your available time.

Don't kill the trees.
Within the Site Planning division, most versions of the vignettes require that candidates avoid disturbing existing trees. Remember that any development that falls under the canopy of a tree will violate this requirement. Also remember that you can add trees to block wind or sun if you are unable to maintain the ideal orientation of your solution.

Use sketch lines.
The ARE software allows you to draw sketch lines while creating your solution. You can use these lines to serve as guides while laying out setbacks, clearances, and element locations before final placement.

Pay close attention to directions.
When reading the program requirements for the Site Planning vignettes, note the difference between words such as "near" versus "adjacent" or "should" versus "shall". In each case, the former gives you more flexibility than the latter.

Align driveways properly.
Keep driveways perpendicular to streets. The site planning vignettes nearly always require perpendicular intersections.

Accuracy and tolerances.
Candidates are responsible for being as accurate as possible when drawing their solution within the graphic divisions, as this results in more accurate scoring. Using the zoom, full screen cursor, and background grid features in the NCARB software will make it easier to produce more accurate solutions. Additionally, a check tool is available in several of the vignettes to identify overlapping elements and other problems.

Although tolerances are built into each scoring program to allow for slight inaccuracies, these tolerances vary from vignette to vignette based on the importance of the feature being evaluated. In general, whenever a specific programmatic requirement is noted in the exam instructions, candidates should be careful to meet that criteria as closely as possible.

ACKNOWLEDGMENTS

This course was written and illustrated by Lester Wertheimer, FAIA. Mr. Wertheimer is a licensed architect in private practice in Los Angeles and a founding partner of Architect Licensing Seminars. For many years he has written and lectured throughout the country on the design aspects of the ARE.

Portions of this edition were revised by Bob J. Wise, Jr., AIA. Bob is a technical director for A. Epstein & Sons International in San Antonio, Texas. He has led ARE review seminars for the San Antonio chapter of the American Institute of Architects for several years, and is a former member of NCARB's exam writing and scoring committees. Bob is also senior lecturer in the School of Architecture at the University of Texas at San Antonio. He led an ARE review session for the Site Planning division at the 2006 AIA national convention.

Special thanks to Thomas Wollan for his contributions to previous updates.

The introduction to this study guide was written by John F. Hardt, AIA. Mr. Hardt is a principal of the firm Andrews Architects, Inc. in Columbus, Ohio. He is a graduate of Ohio State University (MArch) and has been in practice for more than 12 years.

ABOUT KAPLAN

Thank you for choosing Kaplan AEC Education as your source for ARE preparation materials. Whether helping future professors prepare for the GRE or providing tomorrow's doctors the tools they need to pass the MCAT, Kaplan possesses more than 50 years of experience as a global leader in exam prep and educational publishing. It is that experience and that history that Kaplan brings to the world of architectural education, pairing unparalleled resources with acknowledged experts in ARE content areas to bring you the very best in licensure study materials.

Only Kaplan AEC offers a complete catalog of individual products and integrated learning systems to help you pass all nine divisions of the ARE. Kaplan's ARE materials include study guides, mock exams, question-and-answer handbooks, video workshops, and flash cards. Products may be purchased individually or in division-specific learning systems to suit your needs. These systems are designed to help you better focus on essential information for each division, provide flexibility in how you study, and save you money.

To order, please visit *www.KaplanAEC.com* or call (800) 420-1429.

LESSON ONE

THE SITE PLANNING DIVISION

Introduction
The Exam Format
The Site Planning Division
  Division Statement
  Site Design Vignette
  Site Zoning Vignette
  Site Grading Vignette
Emphasis of the Exam
Grading Criteria
The Question of Failure
What Is Expected
How to Prepare
Practice Exams

INTRODUCTION

Site design involves the comprehension, analysis, and design of the elements affecting a parcel of land in order to accomplish some useful purpose. It is a process that results in tangible solutions to practical problems. Site problems may include the placement of structures, development of circulation paths, or design of a surface drainage system. In most cases, site solutions require the modification of existing topography.

Site design has always been an important architectural activity. Throughout history, designers have arranged structures and site elements to achieve practical, secure, and creative solutions to their particular site problems.

THE EXAM FORMAT

The Site Planning division of the ARE consists of three short graphic vignettes. Following is a list of the three vignettes and the amount of time one should allow to solve each of them.

SITE PLANNING DIVISION Total time: 3 hours		
SECTION 1	1½ hours	
	Site Design	1½ hours
Mandatory Break	15 minutes	
SECTION 2	1½ hours	
	Site Zoning	1 hour
	Site Grading	½ hour

Although the time allowed to solve the vignettes totals 3 hours, the appointment time is 3¾ hours, which includes time for general instructions, demographic questions, the scheduled break, and an evaluation that follows the exam.

When a candidate takes the Site Planning exam, the computer randomly selects one version of each vignette type from the pool of available vignettes to create a distinct exam for that individual candidate. Each vignette is designed to be similar and equal in difficulty to all others of its type, so no candidate has an advantage. Therefore, two candidates taking the Site Planning exam at the same time and place will very likely encounter different vignette problems.

Once the preliminary test center procedures are complete, the candidate is instructed to begin the first section. The candidate will have up to 1½ hours to finish the Site Design vignette. It is not recommended that the examinee finish ahead of time. Take the full allotted time to finish each section. Take as much time as needed to review and revise. The first section is complete when the candidate exits the section or when the time limit has expired. At that point the candidate is not permitted to return to that section.

After leaving the first vignette, there is a 15-minute mandatory break. After the break, you may access the second set of vignettes. The procedure for the second set of vignettes is the same as for Site Design, and again, once you exit or the time limit is reached, these vignettes may not be revisited.

THE SITE PLANNING DIVISION

The National Council of Architectural Registration Boards (NCARB) describes the Site Planning division as follows:

Division Statement

The integration of programmatic and site requirements in a responsive and cohesive solution taking into consideration factors such as topography, vegetation, climate, geography, and regulatory aspects of site development.

The division statement continues with a brief description of each of the three vignettes currently offered.

Site Design Vignette

Design a site, including building placement, parking, and vehicular and pedestrian circulation, responding to programmatic, functional, environmental, and setback requirements utilizing general site planning principles.

Site Zoning Vignette

Delineate areas suitable for the construction of buildings and other site improvements responding to regulatory restrictions and programmatic requirements. Define a site profile and maximum buildable envelope based on zoning regulations and environmental constraints.

Site Grading Vignette

Modify a site's topographical characteristics responding to programmatic and regulatory requirements.

EMPHASIS OF THE EXAM

The vignettes described above are designed to test as wide a range of site-related issues as possible. For example, Site Design, the first and longest vignette of this test, requires you to consider the following issues:

- Land use
- Structures placement
- Adjacencies
- Form relationships
- Legal restrictions
- Solar and wind orientation
- Views to and from site
- Pedestrian and vehicular access
- Circulation and parking
- Vegetation and landscaping

You may question how it is possible to deal with all these matters in only an hour, during which you must also read and understand a program with numerous restrictions. First, vignettes are usually uncomplicated problems with few parts. Second, you are not required to produce great architecture, only a relatively mechanical solution to a specific problem.

GRADING CRITERIA

NCARB has stated that most vignettes are designed to allow for many correct answers. The scoring process also allows for minor errors that will not automatically assign a failing score. In a typical vignette, the salient features of a candidate's solution are examined through a complex scoring structure from which a final score is generated. It would be inadvisable and impossible for Kaplan to make suggestions as to what might constitute a mistake that is passable and one that causes a failing score. We instead try to explain the methodology NCARB uses to design and score the ARE.

Consider the following example: You are given a site bordering on a lake, and among the program restrictions is the following: *Provide a 50-foot setback from the lake to any structure.* After you have located the required structures, the program determines whether your design allowed the required 50 feet from the lake. A violation of this requirement would automatically generate an unacceptable score on that aspect of the problem. Similarly, if a 10,000-square-foot pedestrian plaza was required to be located within the building limit lines, the program will know immediately whether the size of the plaza is within the allowable tolerances and whether any part of it encroaches beyond an established limit line.

In a similar manner, hundreds of other significant (and measurable) factors in a vignette are examined, analyzed, and scored. If the preponderance of factors receives an acceptable score, the solution will earn a passing grade.

In establishing scoring standards, NCARB has defined the level of performance necessary to demonstrate entry-level competence. This level is *neither that of a first time employee nor a seasoned practitioner.* It is the level of ability at which an individual can practice architecture independently and without endangering public health, safety, and welfare.

The candidate is well advised to keep in mind that these tests are being graded by computer. The vignettes are, subsequently, designed to test measurable and technical aspects of a problem. The art of architecture is replaced by a series of precise, quantifiable design decisions that are either right or wrong. A building that encroaches on a required setback is unacceptable. NCARB insists that each vignette may have more than one correct solution, but creativity and aesthetics are not part of the equation. Keep foremost in mind that the approach must be to arrange the elements of the vignette in such a way that all the technical and design aspects of the problem are adequately addressed. According to NCARB it is the candidate's responsibility to be as accurate as possible when solving the problems.

> "Basically, more accurate information will result in more accurate scoring"
> (NCARB)

THE QUESTION OF FAILURE

Candidates have always questioned the relatively high failure rate on the Site Planning exam. Since a third or more of all candidates usually fail, some may wonder if an exceptional skill or remarkable performance is required to be successful. If not, then why do so many candidates fail?

There are many reasons why candidates fail this test, but they generally fall into one or more of the following categories:

- Candidates make program-related mistakes. They misread, change, or ignore some programmatic requirement.
- Candidates fail to solve the problem. In some cases, they simply don't know how to solve the problem. At other times, their solutions contain critical mistakes.
- Candidates do not finish. Some run out of time because they take too long to solve the problem. Others may know at once how to solve the problem, but they are insufficiently experienced with the software and its tools to adequately present the solution within the time limit.
- Candidates become bogged down trying to "design" a creative solution, rather than simply solving the problem.

WHAT IS EXPECTED

Candidates for the Site Planning division of the ARE should be aware of what is expected of them. Essentially, you must produce practical solutions within specific established criteria and environments, and present the solutions in a limited time. In all this, one must strive to avoid mistakes that will lower one's grade. For every omission, error, or misjudgment that is identified, a candidate's solution is marked down by the computer program's automatic evaluation. A few trivial mistakes will rarely affect the grade of an otherwise competent solution, but a number of serious errors will likely cause a vignette to fail.

Vignette problems must be solved as simply, accurately, and quickly as possible. Solutions must respond to the site planning issues presented, and the presentation must be clear, direct, and organized. In other words, every aspect of each solution must appear reasonable and professional. If a candidate is able to accomplish these important goals, despite uncertain odds, he or she should be successful.

HOW TO PREPARE

Many candidates are defeated before the exam even begins because their preparation has been inadequate or in some cases, nonexistent. Make no mistake, this graphic exam is tough, and those who believe they can pass it with little or no advance preparation are simply deceiving themselves.

Even if you understand the technical material, the time limit may preclude your completing the vignettes. Therefore, you must be able to apply this knowledge quickly and without hesitation. This takes practice in graphic problem solving, preferably with the kind of exercises that simulate the real exam experience. The more practiced you become in solving graphic site problems, the more rapid, the more accurate, and consequently, the more confident you will become.

The best place to acquire problem-solving experience is in an architectural office. Those who are able to work on all phases of a project will eventually learn about legal site restrictions, building placement, topographic modification, handicapped planning criteria, parking layouts, pedestrian and vehicular circulation standards, etc. Site planning vignettes are intended to be realistic; their purpose is to simulate actual experiences and real problems faced by architects in their daily work. If you are not getting this kind of experience, perhaps you can discuss this with your employer to remedy the situation.

Work experience alone, however, may not provide adequate preparation for all candidates. After all, you could work in an office for years and never be required to design a drainage plan using swales. And solar zoning problems, parking lot layouts, or even a problem requiring the placement of a structure may not come up very often.

Many seemingly obscure site planning problems can be solved by using common sense. In addition, several examples in this course will help you sharpen your skills. The important point is that you must begin somewhere to develop the knowledge, skills, and ability necessary to pass this test. The more extensive your preparation, the better your chance for success.

PRACTICE EXAMS

The Site Planning division does not test design ability as much as it tests your ability to complete all the vignettes in the allotted time. It is an exercise in generating an effective and efficient solution to a problem in a limited amount of time. The experience is deliberately intended to place the candidate under the stress of time pressure. There is simply not enough time to explore alternatives, and therefore, your initial ideas (or their variations) are those that must be developed if you hope to complete the exam within the time limit. Hence, taking practice exams—or mock exams—under the same time constraints and conditions as the real test is the best way to determine design skill.

It is important that the practice problems be similar in scope and scale to those on the actual exam, and they must be solved within the same time limits as the real test. Obviously, given an unlimited amount of time, most candidates would be able to develop an acceptable solution to almost any vignette problem.

Some may wonder if paper-and-pencil exercises can actually help prepare candidates for an exam that is given by a computer. You should understand that the graphic tests require two essential abilities. First, you must develop skill in using the software (covered in Lesson Two). Second, and perhaps more important, you must understand how to solve design problems. Repeated practice, using the practice software furnished by NCARB prior to the exam, will develop the first required skill. The practice software is available at the NCARB Web site. The second skill can be developed through paper-and-pencil exercises such as those found in this book or available from Kaplan AEC Education.

LESSON TWO

THE NCARB SOFTWARE

Introduction
Vignette Screen Layout
Computer Tools

INTRODUCTION

There is a wide variety of drafting programs used by candidates at the firms in which they work. Therefore, an essential part of every candidate's preparation is to practice using the exam software. Candidates may download this software from the NCARB Web site (*www.ncarb.org*). This program contains tutorials and sample vignettes for all the graphic divisions. You are encouraged to spend all the time necessary to become familiar with this material in order to develop the necessary technique and confidence. You must become thoroughly familiar with the software.

The drafting program for the graphic divisions is by no means a sophisticated program. While this may be frustrating to a test taker used to advanced CAD software, it is important to keep in mind that the aim of NCARB in designing these tests was to develop an adequate drafting program that virtually anyone would be able to use—even candidates with no CAD background.

VIGNETTE SCREEN LAYOUT

Each vignette has a number of sections and screens with which the candidate must become familiar. The first screen that appears when the vignette is opened is called the Vignette Index and starts with the Task Information Screen. Listed on this screen are all the components particular to this vignette. Each component opens a new screen when the candidate clicks on it with the mouse. A menu button appears in the upper left-hand corner of any of these screens that returns you to the Index Screen. Also available from the Index Screen is a screen that opens the General Test Directions Screen, which gives the candidate an overview of the procedures for doing the vignettes. Here are the various screens found on the Index Screen:

- **Vignette Directions** (found on all vignettes) Describes the procedure for solving the problem
- **Program** (found on all vignettes) Describes the problem to be solved
- **Tips** (found on all vignettes) Gives advice for approaching the problem and hints about the most useful drafting tools
- **Tree Diagrams** A screen found on the Site Design vignette to show tree shading extents

The beginning of each vignette lesson in this study guide provides a more detailed description of each vignette screen.

To access the actual vignette problem, press the space bar. This screen displays the vignette problem and all the tools required to solve it. Toggle back and forth between the Vignette Screen and one of the screens from the Index Screen at any time by pressing the space bar. This is not as convenient as viewing both the drawing and, say, the printed program adjacent to each other at the same time. Thus it is a procedure that the candidate must become familiar with through practice. Also, some vignettes are too large to be displayed all at once on the screen. In this case, use the scroll bars to move the screen up and down or left and right as needed. The Zoom Tool is also helpful.

COMPUTER TOOLS

There are two categories of computer tools in the ARE graphic divisions.

- Common Tools
- Tools specific to each vignette

The Common Tools, as the name implies, are generally present in all the tests and allow a candidate to draw lines, circles, and rectangles, adjust or move shapes, undo or erase a previously drawn object, and zoom to enlarge objects on the screen. There is also an on-screen calculator, and a tool that lets you erase an entire solution and begin again.

Vignette-specific tools include additional tools that enable the candidate to turn layers on and off or rotate objects. In addition, each vignette also includes specific items under the draw tool required for the vignette, such as property lines, deciduous trees, or grades. Become an expert in the use of each tool.

Each tool is dependent on the mouse. There are no "shortcut" keys on the keyboard. Press the computer tool first to activate it, then select the item or items on the drawing to be affected by the tool, and then re-click the computer tool to finish the operation. Spend as much time as required to become completely familiar with this drafting program. The Common Tools section of the practice vignettes available from NCARB is particularly useful for helping you become familiar with the computer tools. Two things to note: use the left mouse button to activate all tools. Also, there is no zoom wheel on the mouse, nor an associated tool on the program.

The standard computer tools and their functions are shown in Figure 2.1.

FIGURE 2.1 Standard Computer Tools

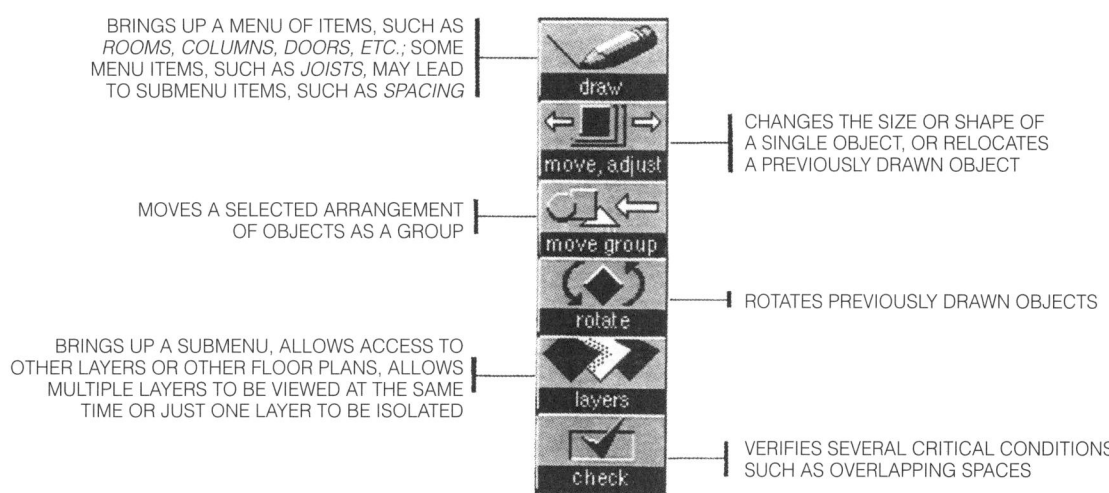

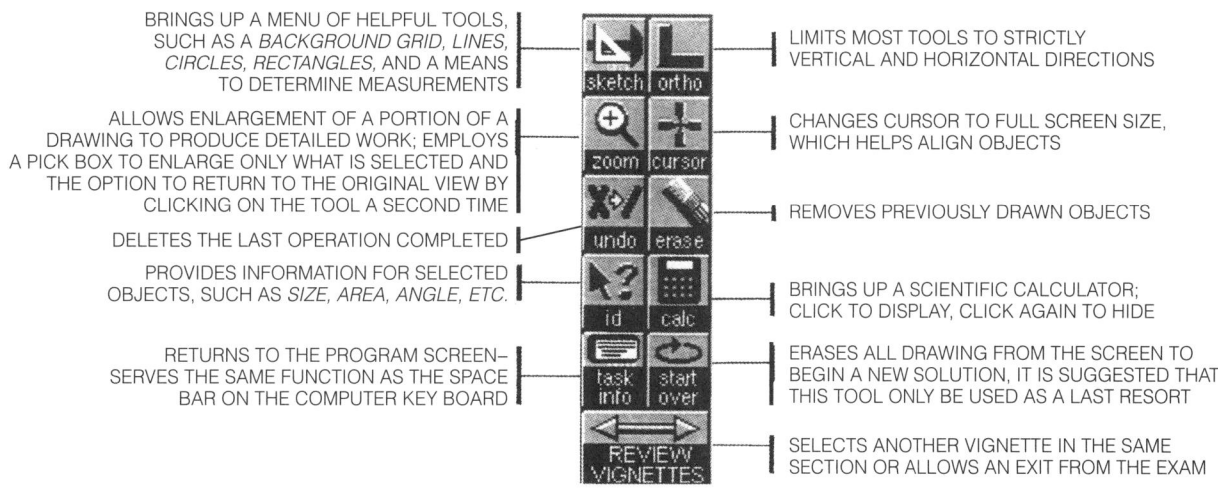

LESSON THREE

TAKING THE EXAM

> Introduction
> Scheduling the Exam
> Final Preparation
> Exam Day
> The Exam Room
> Exam Room Conduct
> Begin with the Program
> General Strategies
> The Time Schedule
> Time Scheduling Problems
> Managing Problems
> Working Under Pressure
> Examination Advice

INTRODUCTION

Preparation for the ARE usually begins several months before taking the actual exam. The first step is to submit an application for registration with your state board or Canadian provincial association. Most, but not all, registration boards require a professional degree in architecture and completion of the Intern-Development Program (IDP) before a candidate is allowed to begin the exam process. Since the processing of educational transcripts and employment verifications may take several weeks, begin this process early. The registration board will review a candidate's application to determine whether he or she meets the eligibility requirements.

SCHEDULING THE EXAM

The graphic exams are available to eligible candidates at virtually any time, since test centers are open nearly every day throughout the year. However, it is the responsibility of the candidate to contact a test center to schedule an appointment. This must be done at least three days prior to the desired appointment time, but it is probably more sensible to make an appointment a month or more in advance. It is not necessary to take the test in the same jurisdiction in which you intend to be registered. Someone in San Francisco, for example, could conceivably combine his or her test-taking with a family visit in Philadelphia.

FINAL PREPARATION

Candidates are advised to complete all preparations the day before their exam appointment, in order to be as relaxed as possible before the upcoming test. Avoid last-minute cramming, which in most cases does more harm than good. The graphic exams not only test design competence, but also physical and emotional endurance. You must be totally prepared for the strenuous day ahead, and that requires plenty of rest and composure.

One of the principal ingredients for success on this exam is confidence. If you have prepared in a reasonable and realistic way, and if you have devoted the necessary time to practice, you should approach the graphic divisions with confidence.

EXAM DAY

Woody Allen once said that a large part of being successful was just showing up. That is certainly true of the licensing examination, where you must not only show up, but also be on time. Get an early start on exam day and arrive at the test center at least 30 minutes before the scheduled test time. Getting an early start enables you to remain in control and maintain a sense of confidence, while arriving late creates unnecessary anxiety. If you arrive 30 minutes late, you may lose your appointment and forfeit the testing fee. Most candidates will begin their test session within one-half hour of the appointment time. You will be asked to provide a picture identification with signature and a second form of identification. For security reasons, you may also have your picture taken.

THE EXAM ROOM

Candidates are not permitted to bring anything (except a calculator) with them into the exam room: no reference materials, no scratch paper, no drawing equipment, no food or drink, no extra sweater, no cell phones, no digital watches. You are permitted to use the restroom or retrieve a sweater from a small locker provided outside the exam room. Each testing center will have its own procedure to follow for such needs. The candidate is allowed to bring his or her own non-programmable, non-printing, non-communicating scientific calculator. The test center staff reserve the right not to permit a calculator if they deem it necessary. Some testing centers may have limited function handheld calculators available. In addition, a calculator is provided as part of the computer drafting program. Scratch paper will be provided by the testing center. The candidate might wish to request graph paper, if available. Not all testing center staff will remember to offer graph paper to ARE candidates.

Once you are seated at an assigned computer workstation and the test begins, you must remain in your seat, except when authorized to leave by test center staff. When the first set of vignettes is completed, or time runs out, there is a mandatory break, during which you must leave the exam room. Photo identification will be required when you reenter the exam room for the next set of vignettes. At the conclusion of the test, staff members will collect all used scratch paper.

Exam room conditions vary considerably. Some rooms have comfortable seats, adequate lighting and ventilation, error-free computers, and a minimum of distractions. The conditions of other rooms, however, leave much to be desired. Unfortunately, there is little a candidate can do about this, unless, of course, his or her computer malfunctions. Staff members will try to rectify any problem within their control.

EXAM ROOM CONDUCT

NCARB has provided a lengthy list of offensive activities that may result in dismissal from a test center. Most candidates need not be concerned about these, but for those who may have entertained any of these fantasies, such conduct includes:

- Giving or receiving help on the test
- Using prohibited aids, such as reference material
- Failing to follow instructions of the test administrator

- Creating a disturbance
- Removing notes or scratch paper from the exam room
- Tampering with a computer
- Taking the test for someone else

BEGIN WITH THE PROGRAM

You can either solve the vignettes in the order they are presented or build confidence by starting with one that looks easier to you. Only you know what works best for you; the practice software should give you a sense of your preferred approach.

Every vignette solution begins with the program. Read the entire program carefully and completely, and consider every requirement. During this review, identify the requirements, restrictions, limitations, code demands, and other critical clues that will influence your solution. Feel free to use scratch paper to jot down key points, data, and requirements. This will help confirm that you understand and meet all the requirements as you develop your solution.

Every vignette problem has two components, the written program and a graphic base plan. Both components are complementary and equally important; together they completely define the problem. For example, on the Schematic Design vignette, review the base plan to identify where users of the building come from and where they will enter the building. Identify the service access or prevailing view or any other essential information mentioned in the program. You should not rush through a review of the program and base drawing in an attempt to begin your design sooner. It is more important to understand every constraint and to be certain that you have not overlooked any significant detail. Until you completely understand the vignette, it is pointless to continue.

GENERAL STRATEGIES

The approach to all vignette solutions is similar: Work quickly and efficiently to produce a solution that satisfies every programmatic requirement. The most important requirements are those that involve compliance with the code, such as life safety, egress, and barrier-free access.

Another important matter is design quality. Strive for an adequate solution that merely solves the problem. Exceptional solutions are not expected, nor are they necessary. You can only pass or fail this test, not win a gold medal. Produce a workable, error-free solution that is good enough to pass.

During the test session, candidates will frequently return to the program to verify element sizes, relationships, and specific restrictions. Always confirm program requirements before completing the vignette, so that you may correct oversights or omissions while there is still time to do so. Candidates must always keep in mind the immutability of the program. That is, you must never—under any circumstances—modify, deviate from, or add anything to the program. Never try to *improve* the program. Only solve what the program asks you to solve, and don't use real world knowledge, such as specific building code requirements.

Candidates should have little trouble understanding a vignette's intent. However, the true meaning of certain details may be ambiguous and open to interpretation. Simply make a reasonable assumption and proceed with the solution.

You'll devise your own strategy for determining in which order complete the vignettes for each section. What follows are some ideas others have found helpful:

1. First, do the vignette that you feel the most capable of completing quickly and accurately. This will be a confidence booster for tackling the remaining vignettes.
2. Try to solve each problem in 10 minutes less than the allotted time. Use this additional time to review your solution.
3. Create a set of notes or a chart for each problem.

THE TIME SCHEDULE

The most critical problem on the exam is *time*, and you must use that fact as the organizing element around which any strategy is based. The use of a schedule is essential. During the preparation period, and especially after taking a mock exam, you should note the approximate amount of time that should be spent on each vignette solution. This information must then become your performance guide, and by following it faithfully, you will automatically establish priorities regarding how your time will be spent.

It is important for a candidate to complete each vignette in approximately the time allotted. You cannot afford to dwell on a minor detail of one vignette while completely ignoring another vignette. Forget the details, do not strive for perfection, and be absolutely certain you finish the test. Even the smallest attempt at solving a vignette will add points to your total score.

Vignettes have been designed so that a reasonable solution for each of the problems can be achieved in approximately the amount of time shown in the *ARE Guidelines*. These time limits are estimates made by those who created this test. In any event, a 45-minute-long vignette may not necessarily take 45 minutes to complete. Some can be completed in 30 minutes, while others may take an hour or longer. The time required depends on the complexity of the problem and your familiarity with the subject matter. Some candidates are more familiar with certain problem types than others, and since candidates' training, experience, and ability vary considerably, adjustments may have to be made to suit individual needs. It should also be noted that within each exam section, the time allotted for two vignettes may be used at your discretion. For example, in a three-vignette section that allots 150 minutes, NCARB recommends spending one hour on one vignette and 45 minutes on the other two. However, you may actually spend 80 minutes on one problem and 35 minutes on the other two.

Candidates who are aware of the time limit are more able to concentrate on the tasks to be performed and the sequence in which they take place. You will also be able to recognize when to begin the next vignette. When the schedule tells you to stop working on one vignette and move on to the next, you will do so, regardless of the unresolved problems that may remain. You may submit an imperfect solution, but you *will* complete the test. Lastly, taking time at the end of each section to review all the vignettes can help to eliminate small errors or omissions that could tip the balance between a passing and failing grade.

TIME SCHEDULING PROBLEMS

It is always possible that a candidate will be unable to complete a certain vignette in the time allotted. What to do in that event? First, avoid this kind of trouble by adhering to a rigid time schedule, regardless of problems that may arise with a particular aspect of the problem. Submit a solution for every vignette, even if some solutions still have problems or are not totally complete.

Candidates are generally able to develop some kind of workable solution in a relatively short time. If each decision is based on a valid assumption and relies on common sense, the major elements should be readily organized into an acceptable functional arrangement. It may not be perfect, and it will certainly not be refined, but it should be good enough to proceed to the next step.

MANAGING PROBLEMS

There are other serious problems that may arise, and while each is potentially fatal, they must be managed and resolved. Consider the following:

- You have inadvertently omitted a major programmed element.
- You have drawn a major element too large or too small.
- You have ignored a critical adjacency or other relationship.

The corrective action for each of these issues will depend on the seriousness of the error and when the mistake is discovered. If there is time, you should rectify the design by returning to the point at which the error occurred and begin again from there. If it is late in the exam and time is running out, there may simply be insufficient time to correct the problem. In that case, continue on with the remainder of the exam and attempt to provide the most accurate solutions for the remaining vignettes. The best strategy, of course, is to avoid critical mistakes in the first place, and those who concentrate and work carefully will do so.

WORKING UNDER PRESSURE

The inflexible time limit of the graphic exams creates subjective as well as real problems. This exam generates a unique psychological pressure that can be harmful to performance. While some designers thrive and do their best work under pressure, others become fearful or agitated under the same conditions. It is perfectly normal to be uneasy about this important event, and although anxiety may be a common reaction, it is still uncomfortable.

Candidates should be aware that pressure is not altogether a negative influence. It may actually heighten awareness and sharpen abilities. In addition, as important as this test may be, failure is not a career-ending event. Furthermore, failure is rarely an accurate measure of design ability; it simply means that you have not yet learned how to pass this difficult exam.

EXAMINATION ADVICE

Following is a short list of suggestions intended to help candidates develop their own strategies and priorities. Each item is important in achieving a passing score.

The *ARE Guidelines,* available from the NCARB Web site, also lists suggestions for examination preparedness.

- **Get an early start.** Begin your preparation early enough to develop a feeling of confidence by the time you take the exam. Arrive at the exam site early and be ready to go when the test begins.
- **Complete all vignettes.** Incomplete solutions risk failure. Complete every problem, even if every detail is not complete or perfect.

- **Don't modify the program.** Never add, change, improve, or omit anything from a program statement. Never assume that there is an error in the program. Verify all requirements to ensure complete compliance with every element of the program. If ambiguities exist in the program, make a reasonable assumption and complete your solution.

- **Develop a reasonable solution.** Since most vignettes generally have one preferred solution, solve the problem in the most direct and reasonable way. Never search for a unique or unconventional solution, because on this exam, creativity is not rewarded.

- **Be aware of time.** The strict time constraint compels you to be a clock-watcher. Never lose sight of how much time you are spending on any one vignette. When it is time to proceed to the next problem, quit and move on to the next vignette.

- **Remain calm.** This may be easier said than done, because this type of experience often creates stress in even the most self-assured candidate. Anxiety is generally related to fear of failure. However, if you are well prepared, this fear may be unrealistic. Furthermore, even if the worst comes to pass and you must repeat a division, all it means is that your architectural license will be delayed for a short period of time.

LESSON FOUR

SITE PLANNING ISSUES: ASSESSING THE SITE

Introduction
- Each Site's Uniqueness
- Manipulating the Landscape
- Visiting the Site
- Checklist

Water
- Introduction
- Uses of Water
- Water in Site Design
- Streams
- Waterfalls and Fountains
- Water Cycle
- Potential Flooding
- Underground Water

Plants
- Introduction
- Defining Space
- Environmental Control
- Aesthetics

Introduction to Topography
- Topography
- Topographic Maps

Contours
- Contour Lines
- Contour Standards
- Contour Patterns
- Contour Characteristics
- Representing Topography

Grading
- Cut and Fill
- Level Areas
- Circulation Paths
- The Grading Process

Drainage
- Storm Drainage Systems
- Drainage Needs
- Surface Drainage
- Sub-Surface Drainage

Design Methods
- Gradients
- Calculating Cut and Fill
- Retaining Walls
- Roads
- Parking Lots
- Building and Topography

Site Analysis
- General Purpose
- Relevant Data
- Climate
- Topography
- Soils
- Hydrology
- Vegetation
- Existing Land Use
- Sensory Qualities
- Natural Hazards

INTRODUCTION

Each Site's Uniqueness

Just as every person has an individual identity, so too each parcel of land has its own unique character. Sites may be large or small, urban or rural—but each one is different, with its own distinctive quality and flavor.

Natural sites often evoke strong emotional responses in us. The starkness of the desert, the grandeur of the mountains, the ruggedness of the coastline—each possesses its own kind of beauty, and each creates a different mood in the observer.

The ability to create a design which evokes the unique characteristic of an area is one of the challenges of an architect. Sustainable Design, encouraging an examination of a broad palette of design components (e.g., environment, urban context, energy systems, landscaping) assists the architect's evaluation of the functional and aesthetic solutions.

Thus, there is an infinite variety of site conditions, and the unique blend of climate, topography, water, soil, rock, and vegetation that form the natural features of a site is not duplicated anywhere else. In this lesson, we will consider these natural site features, except for climate and topography, which are covered elsewhere in this course.

Manipulating the Landscape

If certain sites and site elements are beautiful and functional, while others are not, is it possible for the site planner to manipulate the landscape by removing those elements which are out of character, while enhancing those which are functional and attractive? Yes, often. New planting or other landscape elements can do wonders to heal the wounds caused by neglect or indifference. To be successful, manipulation of the landscape must always be in harmony with the essential character of the land. Does this mean that we should utilize only indigenous materials and vegetation in our site designs? Not necessarily. We simply mean that the structures and plant materials that we introduce and the landforms that we modify should have the same general character as the existing natural site elements, in order to create a pleasing and unified whole.

But some strong natural features cannot be modified. Mountains, rivers, macroclimate—these we must accept and work with, since we are unable to change them.

Visiting the Site

In discussing climate, in the previous lesson, we stressed the importance of visiting the site. Such visits are also vital if the planner is to become fully aware of its other natural features. Only by walking the site can we appreciate the views, landforms, trees, and rock outcroppings that give the site its unique character. But a site is more than the sum of its features; it has an intangible quality—a "feel"—that we try to experience during our visits. Maps and photographs are essential, but there is no substitute for personal observation.

An effective way of recording one's impressions is to take a topographic map or survey into the field and make notes directly on it. Outstanding features, such as trees, rocks, and water bodies are particularly noteworthy, because they are important natural resources, which should be preserved if possible because of their beauty and scarcity. The aesthetic value of such features may even be the dominant reason for selecting a particular site, and the designer must fit his or her designs to them.

Objectionable elements, such as dead trees, weeds, and areas subject to earth movement or flooding are also noted. These one may remove, modify, or avoid, as may be appropriate.

Checklist

The following is an outline of natural site elements to check; of course, not all of them apply to every site.

1. Topography.
2. Climate.
3. Water
    A. Drainage
    B. Surface water bodies
    C. Water quality
    D. Potential flooding
    E. Areas of standing water
    F. Subsurface water conditions
4. Soil
    A. Soil type and bearing capacity
    B. Depth to bedrock
    C. Depth of topsoil
    D. Areas subject to slides or subsidence
    E. Areas of erosion
5. Vegetation
    A. Species
    B. Size, form, density, and condition
    C. Color and texture
6. Views
7. Sounds and smells

WATER

Introduction

Water is essential to all life. Yet in most parts of the United States and other highly developed countries, it is taken for granted, like the air we breathe. We turn on the faucet and it gushes forth; we clean, cook, bathe, flush, and irrigate with it, with little or no thought given to its source or continued availability.

But water is more than a physical necessity—it is a vital part of the landscape both aesthetically and emotionally. Since time immemorial, water has had a tremendous appeal for people. Whatever its form—pool, river, fountain, or waterfall—water is one of the most fascinating of all natural design elements.

Uses of Water

Water has many functions in site design. Some are aesthetic in nature—the still water of a lake is soothing and evokes a feeling of serenity. However, the water body need not be natural to have a strong impact; the rigid geometry of a reflecting pool may also provide a contemplative setting. In contrast to the tranquility of still water, the swift moving water of a fountain or waterfall is dramatic and exciting, both visually and aurally.

Water has a wide variety of practical uses as well. Like all living things, plants and lawns need water to survive and flourish, and in many areas of the country, this is provided by sprinkler systems, which may be designed or specified by the site planner. Water also moderates the microclimate of a site, as discussed in the previous lesson.

The sound of falling water may be used to mask urban noise from cars and other sources, as in Freeway Park in Seattle.

The recreational uses of water are many and varied: swimming in a backyard pool, sailing or waterskiing on a lake, snorkeling in the ocean. The site planner may have to consider the conflicting needs of recreational users, i.e., swimming vs. power boating.

The need for water as recreation may have to be weighed against environmental and other considerations: should a wild river be tamed by damming, or should it be left alone, for scenic enjoyment and whitewater rafting? Some of these decisions are made by the site planner, while others are beyond his or her control. But in any event, the planner should have a thorough understanding of water as a natural resource.

Water in Site Design

Wherever a body of water exists, the land near it is very desirable, and this is reflected in the high price of waterfront property. It seems reasonable, therefore, that any body of water on a site should be preserved, protected, and enhanced. Let's examine each of these goals.

One should preserve a water body by leaving it and the area surrounding it in the natural, undisturbed state whenever possible. We protect it by preventing any kind of contamination. For example, polluted surface runoff should be treated or filtered before being allowed to flow into a body of water. We also maintain natural drainage channels whenever possible and provide detention swales or ponds to prevent flooding.

How can we enhance a body of water? One way is to limit development along the shoreline, thereby creating attractive open space, as well as a much longer effective shoreline set back from the water.

Enhancement of an existing lake may also come about when a dam is built. Dams are constructed for a variety of purposes, including flood control and generation of hydroelectric power. One

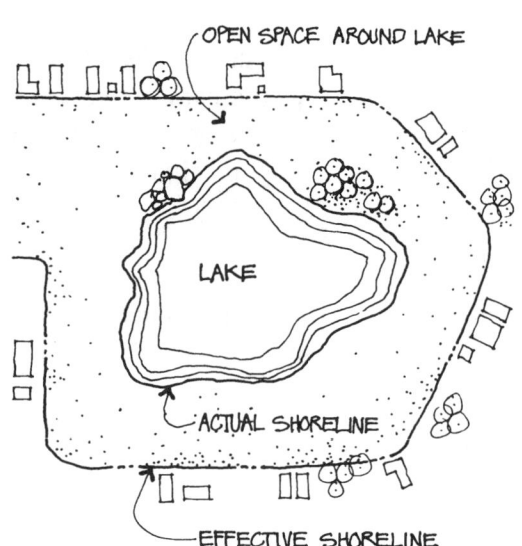

LIMITING SHORELINE DEVELOPMENT

by-product is the creation or enlargement of an upstream lake, which can be used for recreation and as a scenic feature. In recent years, many people have questioned the wisdom of large-scale dam construction, because it inevitably affects, and may even destroy, the natural ecology of an area. Here, as in many other aspects of site and regional planning, one must weigh the advantages and disadvantages of man's intervention in nature.

Just as buildings can become dilapidated through age, use, or misuse, so too some water bodies become unattractive and unappealing: the canals in Venice, California, which were once intended to rival those of their Italian namesake, have fallen into disrepair through neglect. But such water features can be reclaimed and restored.

Where a water body is created or reclaimed, it is often desirable that its shape be natural and curvilinear, rather than artificial and geometric. Of course, this is not always the case. Where man-made forms predominate, as in many urban settings, it is often appropriate for a body of

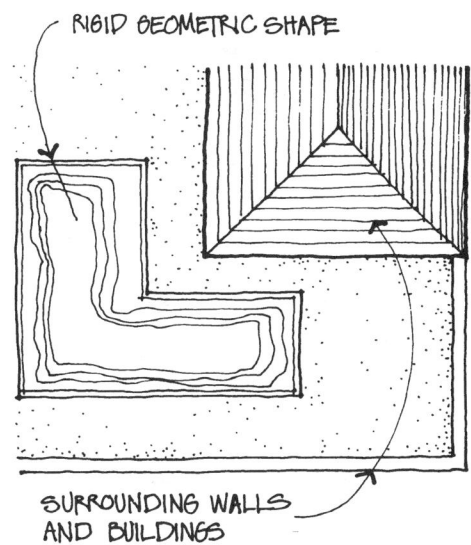

REFLECTION POOL IN URBAN SETTING

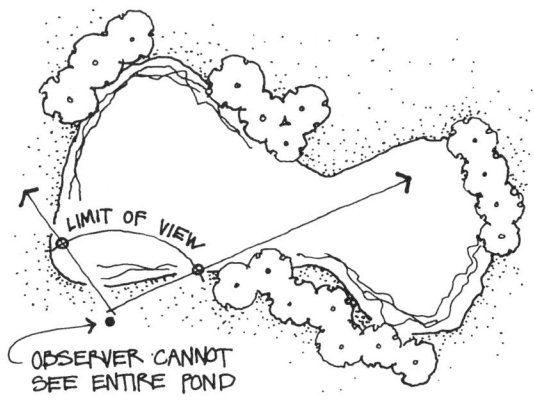

POND IN RURAL SETTING

water introduced into the environment to be rigid in shape and appear man-made.

If possible, the shape of a lake or pond in a rural setting should be such that the entire pond cannot be seen from the shoreline, thus adding to the sense of mystery and the appeal of the pond. Usually, the best sense of balance and harmony is attained when the lake or pond is lower in elevation than any other point in the immediate area.

Paths along the shore should reflect the shape of the pond or lake and the undulation of the water. The design of paths, bridges, docks, and any other structures at the water's edge should be simple and utilize durable and water-resistant materials. Corrosion and weathering are always problems near the water, particularly in the salty atmosphere close to the ocean.

Where possible, banks should be left natural, unless they require surface treatment to withstand the erosive effect of the water.

Such surface treatment may consist of stone, reinforced concrete, treated lumber, or steel, always allowing sufficient freeboard (distance from normal water line to the top of the adjacent surface) for the highest expected water level and maximum wave action.

Whatever the size of the site, the use of water as an outdoor design element adds interest and symbolizes refreshment.

Streams

A stream is any body of water flowing in a channel, such as a river or brook. Its flow varies with the year, the season, and the place, but a stream is always part of a natural drainage system, and therefore, it should be disturbed as little as possible. In general, river banks should also be left alone, because reshaping them or removing existing vegetation may increase erosion.

But nature does not always serve man's needs, and sometimes alterations are necessary. For example, rivers must be crossed. Such crossings should be located where they are most feasible structurally: where the stream is narrow, to minimize the length of span; where the banks are stable, so that economical foundations can be

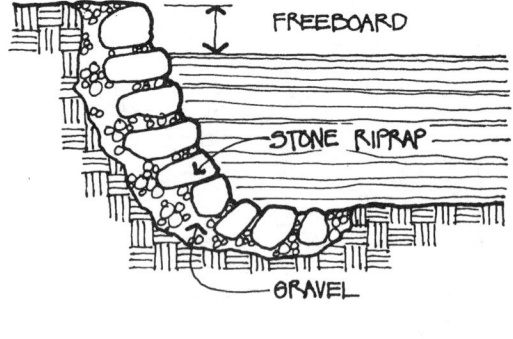

STONE

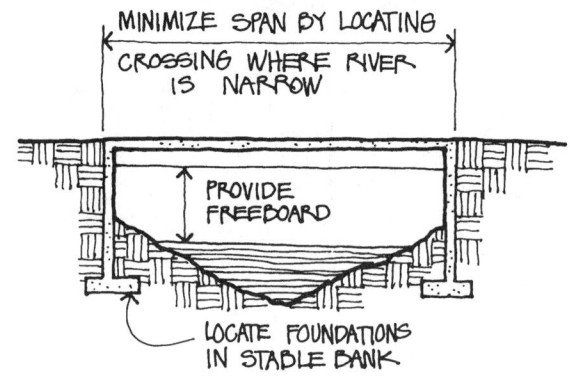

RIVER CROSSING

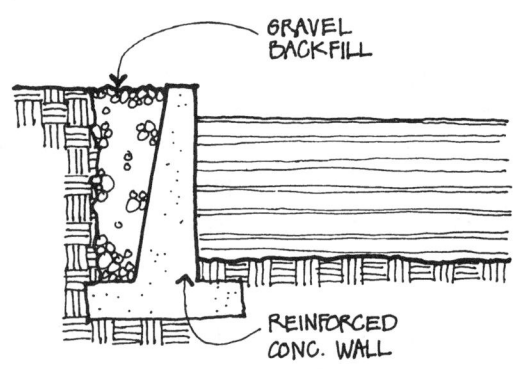

REINFORCED CONCRETE

REINFORCING THE WATER'S EDGE

constructed; and where the banks are higher than the highest expected flood line.

If a crossing must be made in an area subject to flooding, the bridge members should be designed to resist the dynamic action of flood waters.

Where the span between banks is great, additional piers within the stream may be required. Such piers should be oriented with their long dimension parallel to the direction of flow, to minimize disruption of stream flow and the resulting turbulence.

An open man-made drainage channel is also a stream, usually lined with concrete. Such a channel is most efficient if it is straight, without curves or bends, and with a constant width and depth. But a straight channel with a uniform cross-section is not very interesting; a curvilinear channel with a varying cross-section and landscaped banks has a more natural and attractive appearance.

Waterfalls and Fountains

Water falling freely through space because of a sudden change in elevation of its channel creates the most dramatic of all water displays—the waterfall. Natural and man-made waterfalls occur in the outdoor environment in a variety of sizes, shapes, and descriptions. The water may fall in a smooth sheet, or it may be rippled. There may be several falls in segments, or one free fall. And the water may fall into a lake or pool at its base, or onto a hard surface, such as rock.

Each waterfall is unique and creates a different effect. But the interaction of water, light, and sound is always spectacular and always forms a focal point.

The fountain is another dramatic, often theatrical, display of the power of water. Where it is utilized in site design, a large fountain is often the center of attraction, sometimes comprising a variety of jets with multicolored lights and even musical accompaniment. But fountains are not always spectacular—a small, simple fountain, for example, can provide a point of interest in a backyard garden.

Unlike waterfalls, fountains are almost always man-made, with the exception of natural geysers. Regardless of its size or shape, a fountain is perceived as a cool element, making it particularly attractive in a warm, dry climate.

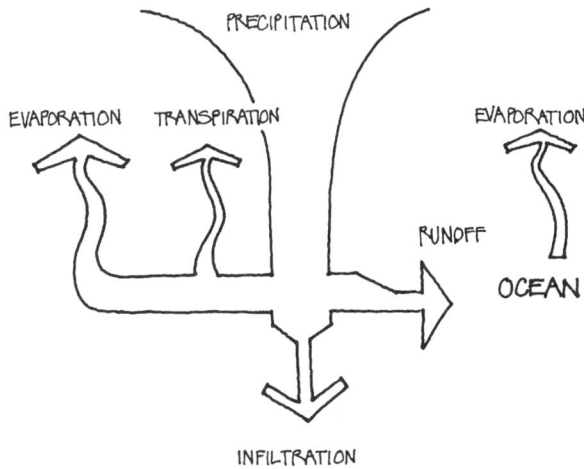

DIAGRAMMATIC WATER CYCLE

Water Cycle

All the water on the earth, under the ground, and in the atmosphere is part of one unified system, called the hydrologic or water cycle. After water falls on the land as *precipitation*, either rain or snow, it follows three paths. A small amount flows off the land into streams and eventually into the ocean as *runoff*. A still smaller quantity soaks into the ground as *infiltration*. Most of the precipitation is evaporated into the atmosphere directly and by *transpiration*, through the action of plants, thus completing the cycle.

Runoff therefore consists of total precipitation, less the water infiltrated into the soil, less the water evaporated directly to the air, less the water transpired back to the air from plants, as shown in the diagram above.

In site planning, we are concerned with all of these hydrologic processes, particularly runoff and infiltration. When a site is developed, the amount of runoff increases. To understand why this is true, let's look at the water cycle diagram. Site development generally entails the removal of some vegetation, thus decreasing the amount of transpiration. Also, relatively pervious land is replaced by impervious buildings, streets, and parking areas, which reduces infiltration. Less transpiration into the air and less infiltration into the soil mean more surface runoff.

How does the site drainage system handle this surface runoff? One approach is for the runoff to immediately enter the drainage system. A different approach requires that most rainfall be held in a *detention pond* on the site until the rain subsides. The runoff is then released slowly without causing flooding. The intent of this second approach, which is often required by local ordinance, is that the flow of rain water from the new development be equal to the runoff from the site prior to development. Sustainable Design would suggest that the second approach is more environmentally appropriate. In fact, it would encourage the slopes of the ponds to be covered in wetland vegetation to allow the entering waters to be naturally filtered.

Potential Flooding

We have said that the development of a site increases the amount of runoff. On a larger

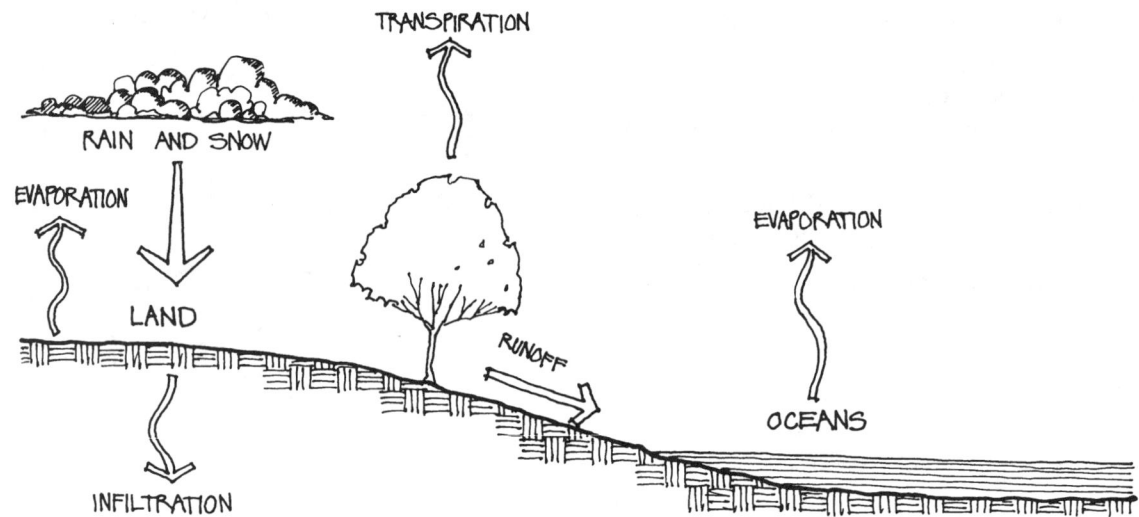

THE WATER CYCLE

scale, urbanization has a similar effect on the hydrologic cycle: the amount and speed of runoff are increased, the runoff is warmer and contains pollutants, and the stream which eventually carries the runoff is visually impaired as a result of erosion and pollution. Urbanization in this country has been especially rapid during the past 30 years. Consequently, there have been a number of disastrous floods. Although measures have been taken by various public agencies, the potential danger of flooding remains high in some areas.

The relatively flat land within which a stream flows is called a *flood plain*. When the volume of flow exceeds the stream's capacity, it overflows its banks and spreads onto the flood plain. This occurs regularly, and the boundary of the flooded area depends on the flood frequency. Thus a ten-year flood inundates less land than a 100-year flood.

The term ten-year flood refers to a flood of a magnitude such that it is likely to occur only once every ten years; therefore, the likelihood of a ten-year flood occurring in any given year is ten percent. Similarly, a 100-year flood is one that is expected to occur once every 100 years. A 100-year flood therefore has a much greater magnitude than a ten-year flood. When designing in flood-prone areas, we select a flood of a given magnitude—say a 100-year flood—and set floor elevations above that flood level.

Since the flood plain is subject to natural and recurring floods, it seems obvious that any construction within the flood plain courts disaster. Therefore, such lands should be limited to open space uses, such as recreation and agriculture.

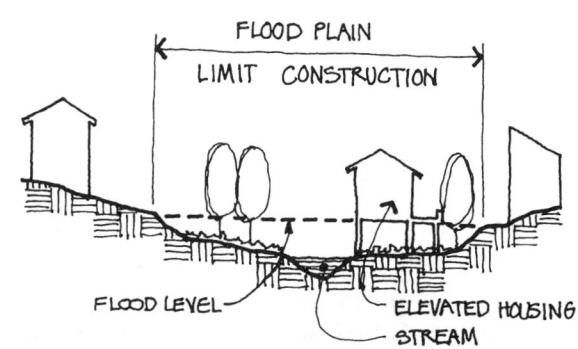

SECTION THROUGH FLOOD PLAIN

This swath of land can provide a natural, park-like, easily-maintained setting. Unfortunately, it doesn't always work out that way; as desirable land for development becomes scarcer and more expensive, there is increased pressure to build on flood plains. As a compromise, low-density housing is often permitted, provided the occupants are aware of the potential hazard and the structures are elevated above flood level.

The water table in a flood plain usually occurs near the surface, drainage is generally poor, and the soil deep and uniform. The soil is often subject to large volumetric changes when it becomes wet, making it unsatisfactory for supporting building loads, but usually excellent for agriculture. The rivers in flood plains are often meandering.

The conventional solution to the problem of potential flooding involves the construction of concrete channels. Alternatively, existing natural drainage channels and flood plains can be utilized, even if some modifications are required. Such solutions preserve the aesthetics and ecology of the natural environment and are compatible with the Sustainable Design philosophy.

Underground Water

The water contained in the voids and crevices under the earth's surface exceeds by far all of the water contained in streams and lakes. This underground water comes from precipitation, both rain and snow, which soaks directly into the ground or drains into rivers and lakes and then seeps into the ground.

Underground water is found either in the zone of aeration or the zone of saturation. The zone of aeration is the higher zone, where the spaces between the soil grains contain both water and air. In the lower zone, the zone of saturation, all

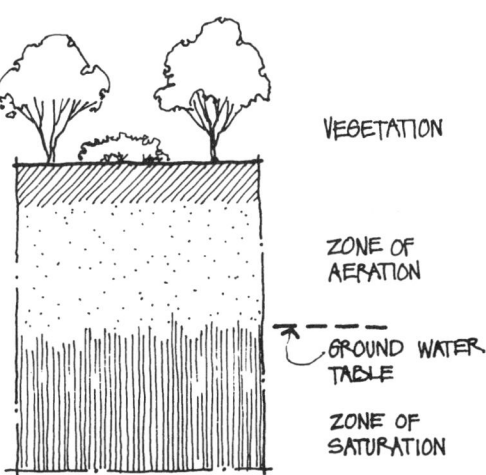

UNDERGROUND WATER ZONES

of the void spaces are filled completely with ground water. The irregular surface which forms the boundary between the two zones is called the *ground water table*. This is usually a sloping surface which fluctuates seasonally and roughly follows the ground surface.

Where the ground water table is high, about six feet below the surface, construction excavation must be braced and kept dry by pumping. Basements must be waterproofed, basement walls designed to resist hydrostatic pressure, underground tanks or other structures designed to resist uplift, and the bearing capacity of the foundation soils often reduced.

Underground water flows at a very slow velocity, depending on the porosity and permeability of the underground earth or rock material. An underground permeable material through which water flows is called an *aquifer*. Sand, gravel, sandstone, and some limestones are generally good aquifers, while clay, shale, and most metamorphic and igneous rocks are poor aquifers.

PLANTS

Introduction

Plants are an important site design element and provide beauty and vitality to the outdoor environment. A site development without plants would be like a moonscape—stark and lifeless. Plants soften the hard edges, define spaces, and add interest.

The use of planting in site design is not merely ornamental or decorative, any more than a building design is. Rather, the site planner often considers planting as a functional element and an integral part of the overall design of the site. In fact, indigenous planting is usually easier to maintain and often provides functional benefits such as filtering storm water or providing habitat for local wildlife.

Plants are unique in that they are the only design element that is alive and hence ever-changing. They grow, change with the seasons, and move with the winds. The site planner must be aware of seasonal characteristics and growth patterns, and realize that the plant that is placed in the ground today will look different next month and next year.

Other design elements can often be ignored after a site is developed, but plants, being alive, need constant nurturing; they need the right soil, sun and wind exposure, temperature range, moisture, and nutrients to live and thrive. The site planner considers this need for ongoing maintenance and often selects plant materials that are relatively easy to care for. On the other hand, where proper maintenance of vegetation is simply too difficult or expensive to achieve, other elements, sometimes inferior, are used as substitutes—for example, blacktop paving instead of grass in a school yard.

Like all natural things, plants are imperfect and not totally predictable. That is the essence of their appeal: they provide a connection with nature.

Defining Space

When we speak of plants or vegetation in this lesson, we are referring to all living organisms in the environment which draw their sustenance from the soil. Trees, shrubs, ground cover, lawn—all are plants. Our discussion comprises both indigenous vegetation and plants introduced into the environment. Since native plants, by their very existence, are well suited to a site, they should be preserved, unless there is an overriding reason to remove them.

New plants should not be introduced haphazardly, but only after careful consideration. A well thought out landscape can change an ordinary site into one that is attractive and distinctive.

In addition to their aesthetic value, plants serve a variety of other functions in the outdoor environment, including defining space. In a building, space is usually defined by rigid physical elements, such as walls, ceiling, and floor. Outside, the definition of space is more subtle. Trees

TREES CREATE HORIZONTAL ENCLOSURE

or shrubs may provide a feeling of vertical enclosure, without actually enclosing an area. With deciduous trees, the spatial definition is much stronger in the summer than in the winter, when the trees have lost their foliage. Closely-spaced trees may also provide a horizontal enclosure, or ceiling.

In addition to forming enclosures, trees may visually connect structural elements, such as buildings, and direct people into a space. Thus, the site planner can create a variety of spatial feelings through the use of plants.

Plants can act as a screening device; a cluster of tall, dense trees may provide privacy for an outdoor terrace, for example.

While the trees block the view into the terrace, they also prevent viewing from the terrace, and thus separate it from its surroundings.

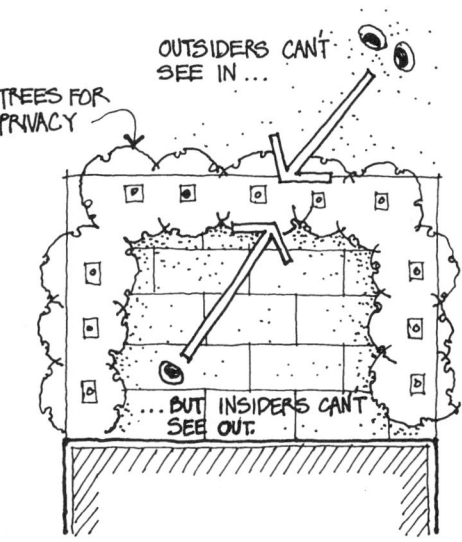

PRIVACY CREATES SEPARATION

Unattractive site elements may be visually screened by planting: most people would rather look at trees than mechanical equipment or parked cars.

Of course, planting used for privacy or screening should be evergreen, in order to be effective throughout the year.

Environmental Control

Vegetation is one of the most moderating influences on the environment. Trees block both the sun and the wind. They act as nature's air conditioning by cooling, humidifying, and filtering the air. They create sheltered zones by reducing wind speeds.

Trees and other planting help to control erosion, destructive runoff, and flooding. They absorb sound. And they provide a habitat for birds.

In these and other ways, plants improve the quality of the environment and hence, the quality of human life.

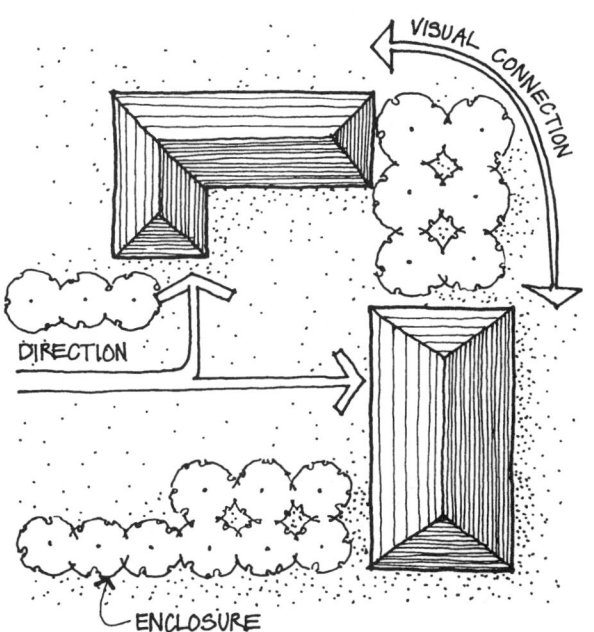

TREES ENCLOSE, DIRECT, AND CONNECT

Aesthetics

For convenience, we have separated the various functions of plants. In reality, of course, all of these functions may be performed simultaneously; a tree can provide shade, help to define space, and look beautiful, all at the same time.

Trees are the dominant plant material; they must be carefully selected and placed to assure a successful design. In general, trees should be clustered as they are in nature, and not spaced too regularly or too far apart. A row of uniformly spaced trees tends to look formal and unnatural; however, it may be appropriate along a street or in an urban setting, to reflect the rigid forms of the built environment.

Occasionally, a single tree can be used as a focal point in the outdoor environment, in much the same way as a piece of sculpture. It may stand by itself, or it may be complemented by smaller trees and plants to create a unified composition.

While the larger trees are dominant, smaller trees and shrubs are used to subdivide the site into smaller areas, visually connect the various site elements, define paths or roads, and add visual interest.

If we think of larger plant materials forming the walls and ceiling of the outside environment, then ground cover is its carpet. Ground cover defines a space or surface, provides visual interest because of its color or texture, and retains soil and moisture.

Trees or other plant materials may be used to frame a view.

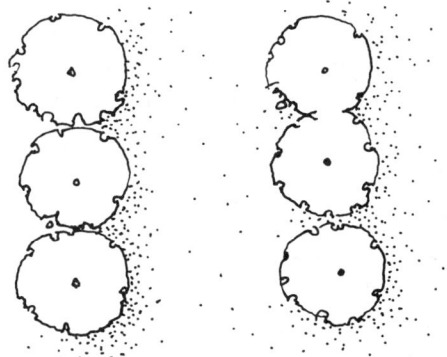

INFORMAL & NATURAL

FORMAL & UNNATURAL

TREE GROUPINGS

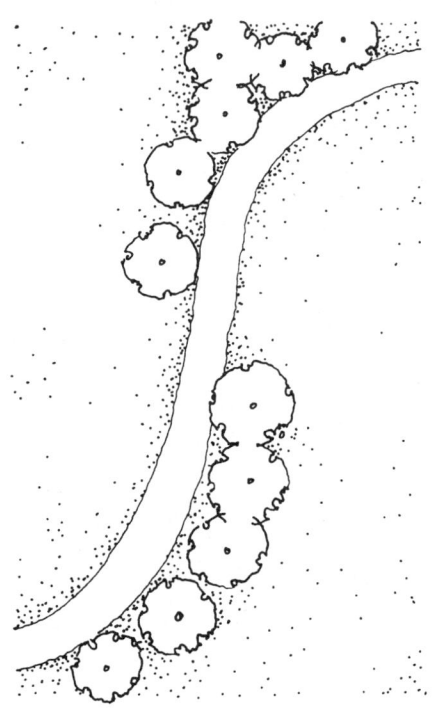

SHRUBS DEFINE A PATH

Just as the facade of a building should be free from clutter, without too many materials or fussy details, so too one should avoid the use of numerous plant varieties and complex groupings. It's best to keep it simple, using just a few carefully chosen, grouped, and placed plant varieties.

However, this does not mean that only one plant size or form should be used. There is an almost infinite variety of sizes, forms, textures, and colors of plants to choose from. In general, most of the plants in a site design should be more or less conventional and appropriate to their surroundings. However, plants of varied shapes, colors, and textures may be added to provide more interest. Too little variation is dull and monotonous, while too much is busy and even chaotic.

INTRODUCTION TO TOPOGRAPHY

Topography

Topography is the graphic representation of an area's surface features. It is synonymous with landform or the shape of the ground. It may encompass mountains, rolling hills, prairies, and plains, while at a different scale, it may include mounds, ramps, berms, and even ripples in a sand dune. Topography has great environmental significance, since it affects the aesthetic character of an area, its microclimate, drainage, views, and the setting for structures.

Since the ground surface of every area has a unique configuration, the topographic map of a particular site will be unique as well. A topographic map allows one to understand the pattern of the land, since it indicates slopes, ridges, valleys, summits, stream beds, and drainage patterns. Topographic maps, therefore, are an invaluable aid for environmental assessment; in fact, topography is often the major determinant in site development.

Topographic information always influences and frequently determines site use, site circulation, distribution of utilities, placement of buildings, and the disposition of open spaces. A school playground, for example, requires a large, relatively flat area; a dead level area, on the other hand, may have drainage problems. Without a topographic map for reference, slope analysis for such problems would be impossible. Site design, therefore, is largely dependent on landform information provided by topographic maps.

Topographic Maps

Topographic maps are developed either from aerial photographs or surveys, where smaller parcels are involved. Aerial photos have the inherent advantage of depicting the land with a high degree of detail. Individual photos can be interpreted separately; however, they are most useful when combined in pairs and viewed stereoscopically. With the use of special equipment, aerial photos may be scanned to determine lines of equal ground elevation, that is, contours.

Topographic surveying consists of obtaining field data, from which a map is plotted showing

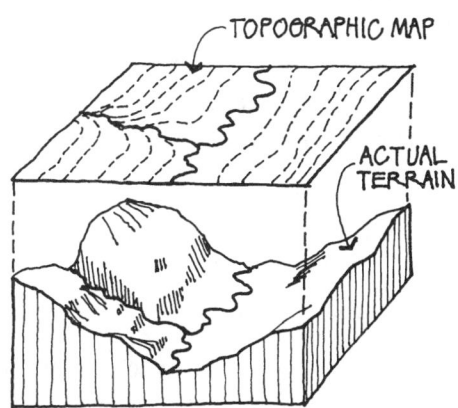

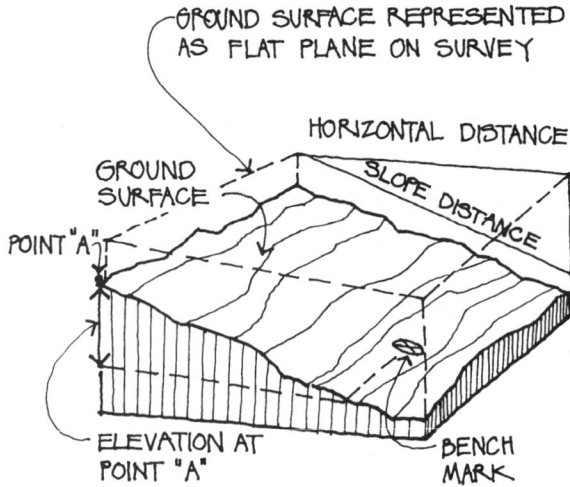

the configuration of the ground surface. Regardless of the method used, the purpose of all topographic surveys is to produce a map on which variations in ground elevation can be readily observed and analyzed. Topographic maps also generally show property lines, roads, structures, trees, etc., in addition to ground surface elevations.

Surveys that extend over a relatively small area ignore the earth's curvature and assume the ground surface to be a flat plane. Horizontal distances, therefore, are considered to be straight and are measured along a flat plane. Slope distances are never used for measurement. Vertical distances, or elevations, are designated as the distance above sea level or above any other bench mark, that is, any permanent point of known elevation. Both horizontal distances and elevations are measured and recorded in feet and hundredths of a foot, never in inches.

CONTOURS

Contour Lines

The shape of the ground surface is most often represented on drawings by contour lines. Contours are imaginary lines that connect all points of equal elevation. Each contour line may be thought of as the intersection of a level plane with the ground surface, such as a horizontal slice through a mountain or the shore line of a lake. Contours were developed in Holland in the mid-18th century, but it was not until the late 19th century that they became the common method for depicting terrain on survey maps.

The contour interval is the uniform difference in elevation between two adjacent contours. This interval is typically 1, 2, 5, or 10 feet, depending on the purpose and scale of the map and the character of the terrain represented. The contour interval is generally constant for the entire drawing and should always be indicated somewhere on the plan.

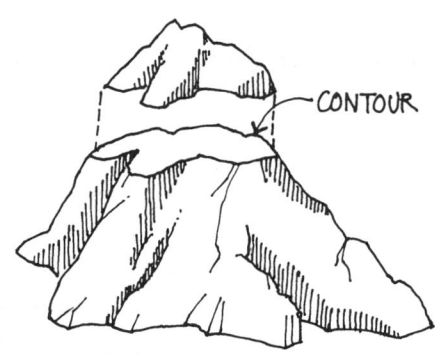

HORIZONTAL SLICE THROUGH A MOUNTAIN

SHORELINE IS ALSO A CONTOUR LINE

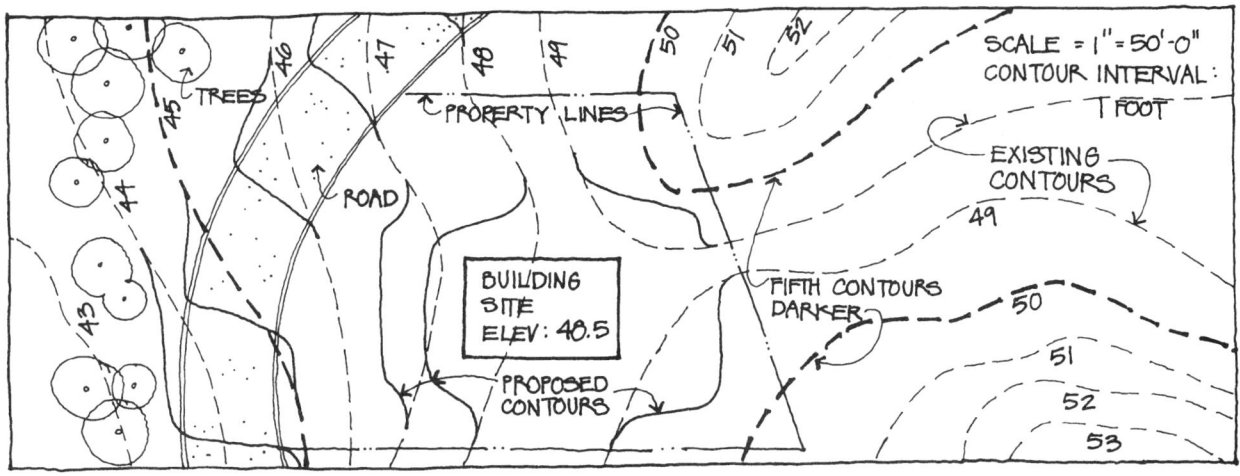

CONTOUR STANDARDS

Contour Standards

Following are some standards that apply to the use of contours:

1. Existing contours are shown by a dashed line.
2. Every fifth contour is shown slightly darker for legibility.
3. Proposed (or modified) contours are shown by a solid line on the same drawing that shows existing contours.
4. Contour lines are labeled with the number within or on the high side of the line.
5. The contour interval is small for relatively flat areas, while for rough terrain, the contour interval is large.
6. The smaller the scale of the map, the larger the contour interval.

Contour Patterns

For those who work with topographic maps, it is important to know the contour patterns of typical topographic features. Once these are mastered, it becomes a simple matter to understand the shape of a ground area and to readily modify landforms when required. Following are the most typical topographic forms:

1. *Uniform slopes* are indicated by parallel contours which are evenly spaced.

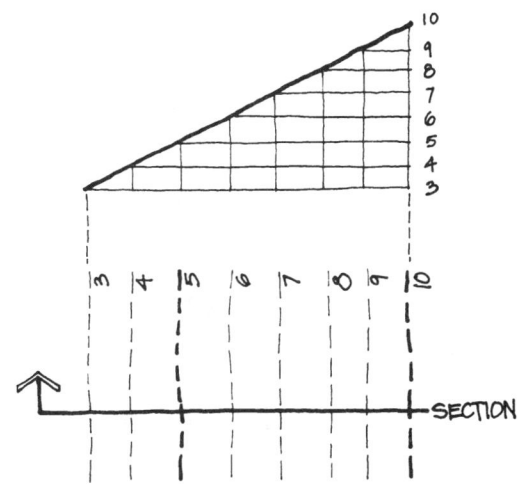

1- UNIFORM SLOPE

2. *Convex slopes* are shown by parallel contours spaced at increasing intervals going uphill. In other words, the closer contours are at the lower elevations.

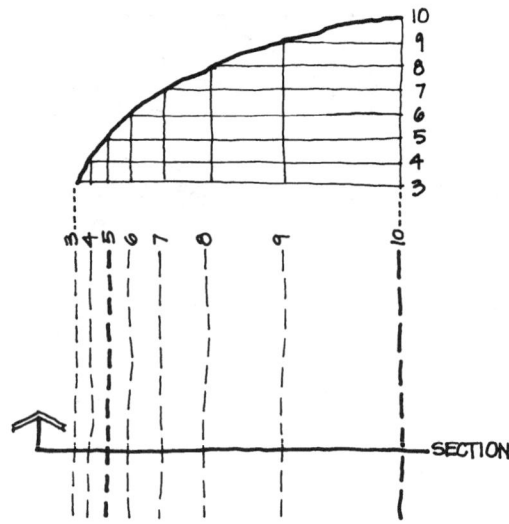

2 - CONVEX SLOPE

3. *Concave slopes* are shown by parallel contours spaced at decreasing intervals going uphill. In this case, the closer contours are at the higher elevations.

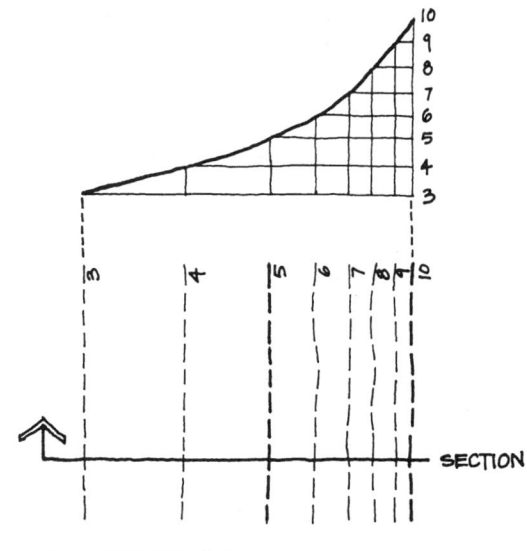

3 - CONCAVE SLOPE

4. *Valleys* are indicated by contours which point uphill.

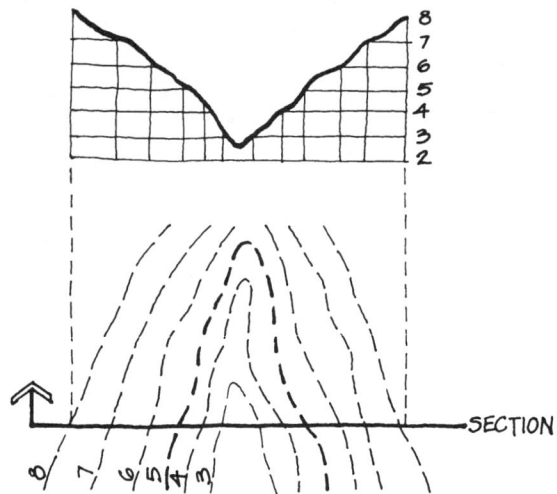

4 - VALLEY

5. *Ridges* are indicated by contours which point downhill.

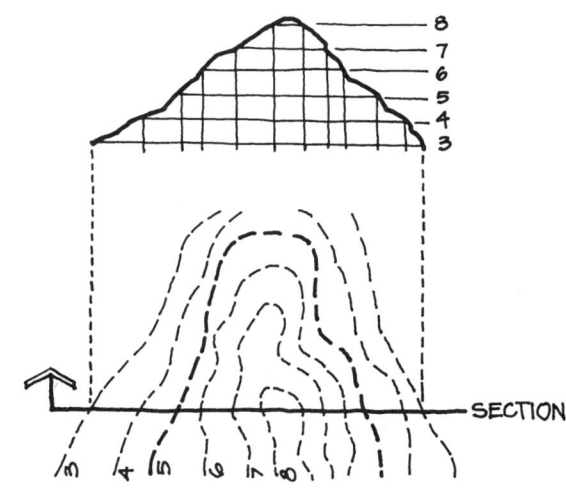

5 - RIDGE

6. *Summits* and *depressions* are represented by concentric closed contours. For both forms, spot elevations should be included at the highest or lowest point.

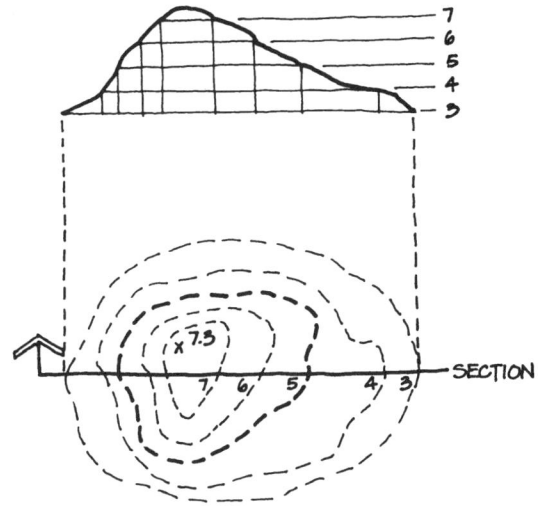

6 - SUMMIT

Contour Characteristics

Following is a list of the general characteristics of contours. The letters refer to the map on page 31.

A. All points on a contour have the same elevation.
B. Contour lines never split, although two identically numbered contours may appear side by side, at the top of a ridge or bottom of a valley.
C. Contour lines never cross, except where there is an overhanging cliff, a cave, or similar configuration.
D. Equally spaced contours indicate a uniform sloping surface.
E. Contours spaced close together indicate a steep slope.
F. Contours spaced far apart indicate a slight grade.
G. Contours spaced at increasing intervals (further apart) going uphill indicate a convex slope.
H. Contours spaced at decreasing intervals (closer together) going uphill indicate a concave slope.
I. Valleys are indicated by contours pointing uphill.
J. Ridges are indicated by contours pointing downhill.
K. A contour that closes on itself within the map area is either a summit or a depression.
L. Contours that run in straight parallel lines indicate a plane surface.
M. Drainage always occurs perpendicular to the contours, because this is the shortest distance and hence the steepest route of travel.

Representing Topography

Although contours are the most common way of representing topography, there are other methods used to depict topographic relief. Among these are the following:

Spot elevations. A spot elevation is a number corresponding to the exact elevation at a key point on the ground. Spot elevations are designated by an arrow pointing to the exact spot where the elevation is located, and they are used to indicate high or low points, tops of curbs, bottoms of walls, bases of trees, floor levels of structures, building corners, etc. Spot elevations may also be used on a uniform grid with precise elevations noted at every intersection point of the grid. Contours may be determined from spot elevation grids through interpolation.

Shading. There are several shading methods one can use to depict ground form, and all of them involve shading the slopes in proportion to their degree of steepness. Shallow slopes are lighter with the shading lines further apart, while steep slopes are made darker by placing the shading lines closer together. The most common

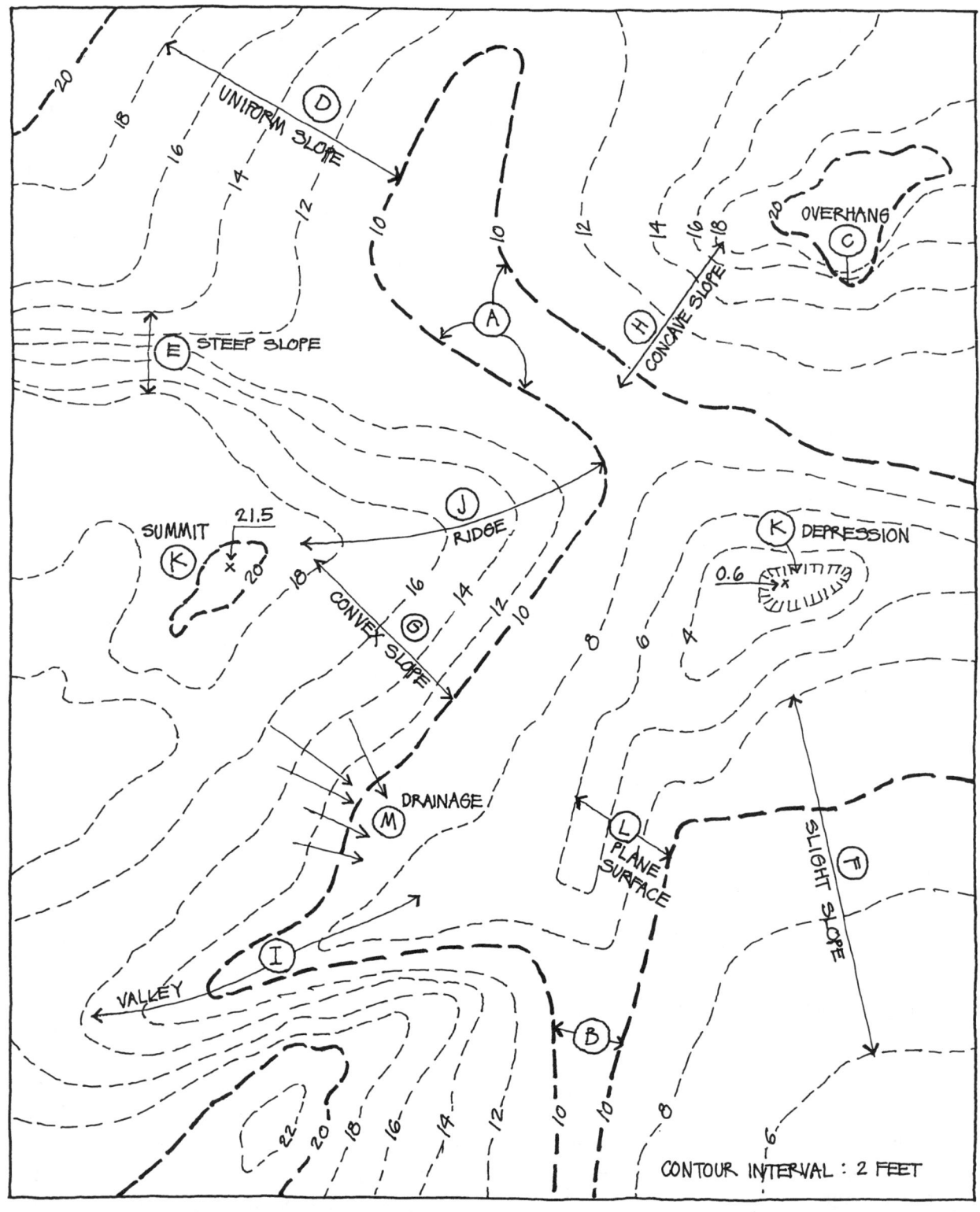

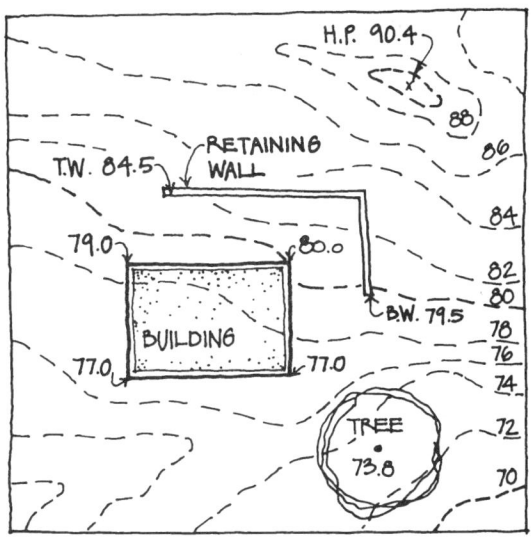

SPOT ELEVATIONS

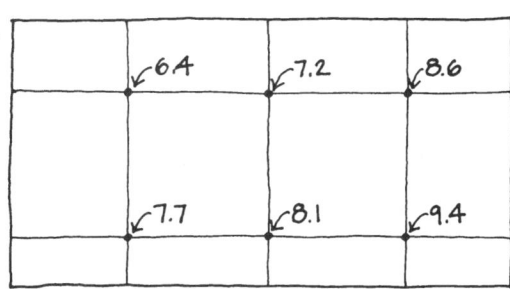

SPOT ELEVATIONS ON GRID

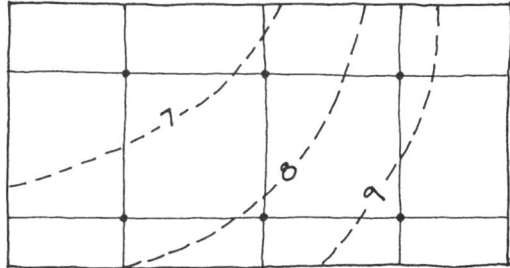

CONTOURS INTERPOLATED FROM SPOT ELEVATIONS

shading system employs *hachures*, which are short, disconnected lines drawn perpendicular to the slope, or in the direction of water flow. Hachuring requires knowledge of the area's contours, since the shading lines are drawn

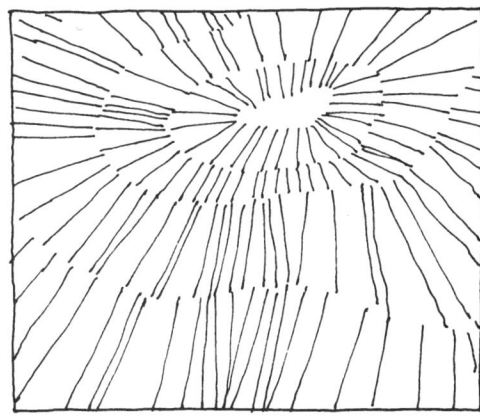

HACHURING

between two successive contour lines. Hachured maps give a vivid picture of the ground terrain, but it is difficult to determine exact elevations unless spot grades are indicated. For this reason, hachuring is rarely used today.

Models. One of the best ways of studying the general form of the terrain is through the use of relief scale models. Contour plans can be converted into useful three-dimensional models by transferring the contours onto a layer of material (cardboard, polystyrene, etc.) whose thickness represents the contour interval, and then gluing the layers together. Rough models are quickly and easily made, and they enable the viewer to comprehend the ground form of an entire site at a glance.

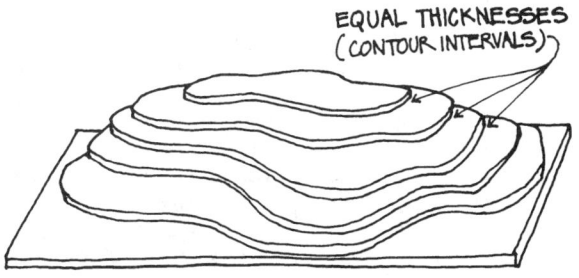

RELIEF SCALE MODEL

GRADING

Cut and Fill

The development of a site almost always requires a certain amount of topographic modification, that is, grading. In general, Sustainable Design encourages grading that should be kept to a minimum, so as not to upset existing drainage patterns, disturb native landscaping, or destroy an area's natural aesthetic value.

There are only two ways to change the land surface: by removing earth (cutting) or by adding earth (filling). The greatest economy results when the amount of earth cut is approximately equal to the amount filled. Grading plans indicate cut or fill by means of new, solid contour lines, which represent proposed modifications to the existing contours, which are shown dashed. A proposed contour that moves in the direction of a lower contour line indicates fill, that is, the addition of earth. Conversely, a proposed contour that moves in the direction of a higher contour line indicates cut, or earth removed. All grading involves cutting, filling, or a combination of the two.

Level Areas

Almost all structures require a fairly level ground area on which to sit. Creating a level area, therefore, represents one of the most common exercises in modifying contours. A familiarity with these basic methods will enable one to understand the essential principles of modifying all ground form.

A level area can be created in one of the following three ways:

1. By cutting into the slope
2. By filling out from the slope
3. By a combination of cutting and filling

Cutting into the slope provides a level area by essentially scooping it out of the hillside. You begin by locating the area to be leveled on the topographic map. The elevation of the contour just downhill of the level area will be the finish grade for the entire level area. Next, you take the nearest higher contour and wrap it around the high side of the level area and connect it to the same existing contour. This first modified contour should be located a few feet away from the level area to allow for the ground slope. For example, if the contour interval is one foot and the finish grade of the slope is to be 2:1 (2 feet horizontal to 1 foot vertical), then the first modified contour should be drawn two feet away from the level area. Proceed now to the next higher contour and wrap it around the level area in the same way, keeping the same distance between contours, that is, two feet for a 2:1 slope. Continue up the hillside until there are no more overlapping contours, being certain that

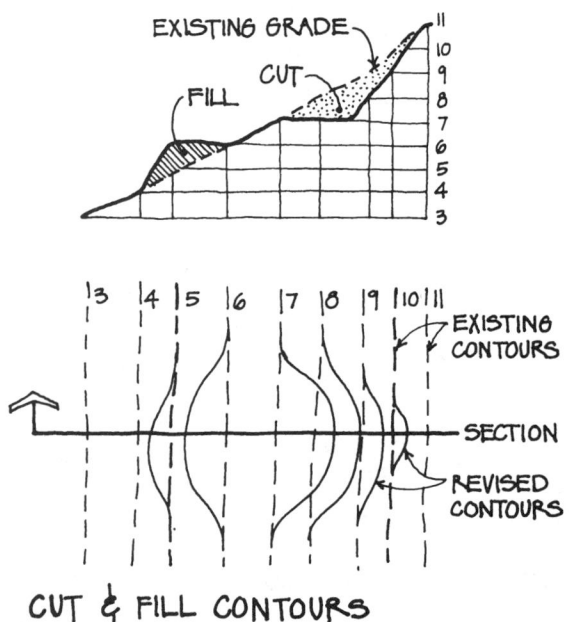

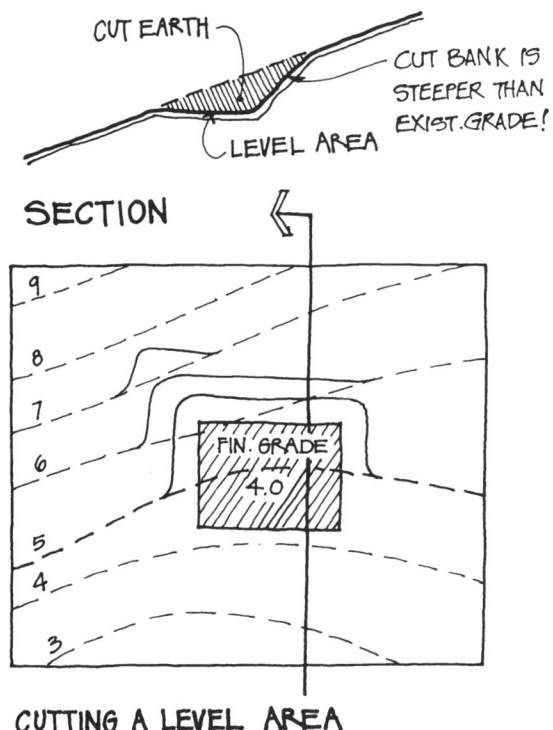

CUTTING A LEVEL AREA

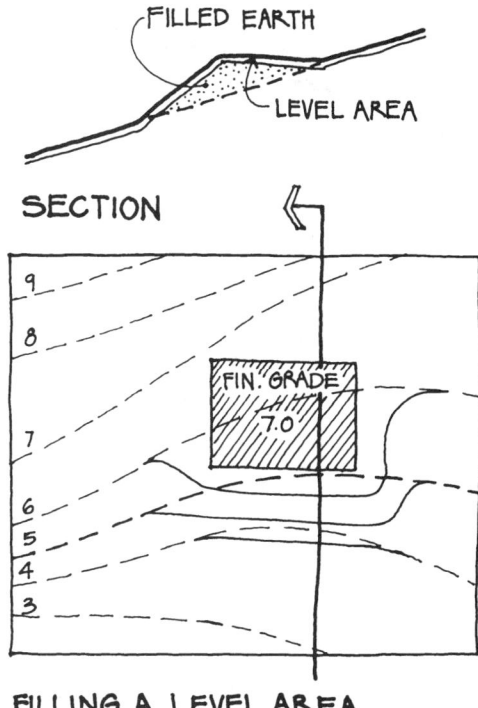

FILLING A LEVEL AREA

the two-foot minimum distance between contours is maintained.

It is important to select a gradient for the new sloped bank that is steeper than the existing grade; otherwise you will never be able to meet the existing grade, that is, the excavation of earth will never end. Normally, one attempts to minimize grading by meeting the existing grade as quickly as possible, provided that the new bank is a stable slope. With steeper slopes, vertical cuts and retaining walls may be necessary.

Filling out from the slope is substantially the reverse of the method just described. Instead of scooping out the hillside, you create a level shelf by adding earth. Again you begin by locating on the topo map the area to be leveled. The finish grade of this area is set at the same elevation as the contour immediately uphill of the level area. Next, you take the nearest lower contour and wrap it around the low side of the level area. As before, the contour is spaced far enough away from the level area to accommodate the desired slope between contour lines. You continue downhill, wrapping contours around the level area, until your new slope angle meets the existing grade.

A combination of cutting and filling is the most common grading method for providing a level area, since this method attempts to balance cut and fill. If the amount of earth cut is equal to that used for fill, the need to import or dispose of soil is eliminated and grading costs are reduced to a minimum. After locating the area to be leveled, you establish the area's finish grade by determining the existing grade at its midpoint, which is at the intersection of the level area's diagonals. Next, you wrap the adjacent contour above the level area around the high side, and the next lower contour around the low side. This same procedure is followed, as before, until all contour lines requiring modification have been adjusted.

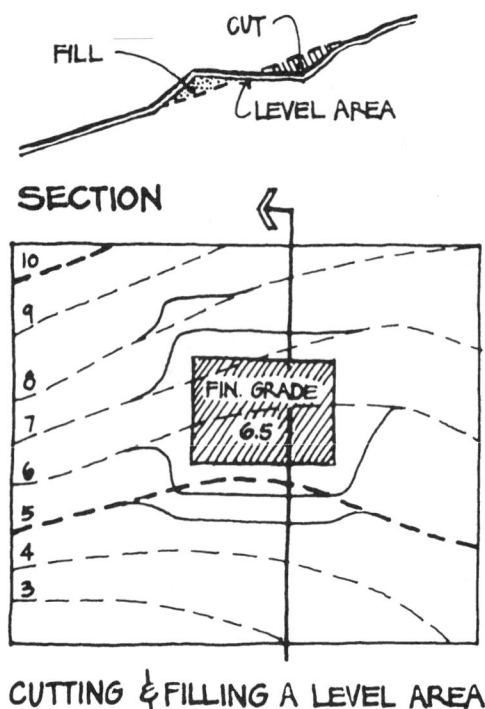

CUTTING & FILLING A LEVEL AREA

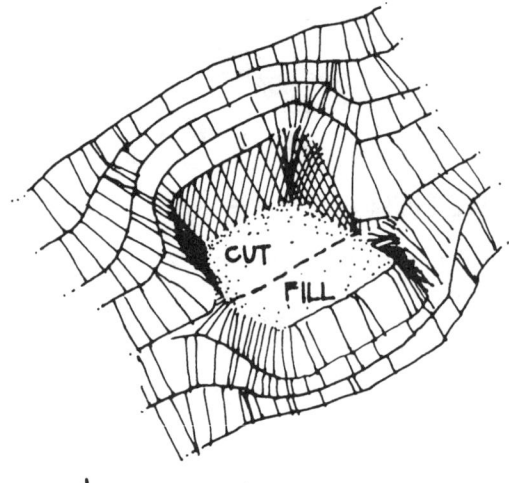

CUT & FILLED LEVEL AREA

Given a choice, which grading method should be used? Often, the choice is out of the designer's hands. For example, filling operations may not be feasible on steep sites, in which case cutting may be the only solution. But, cut earth must be hauled away, which is expensive. Cut earth is generally more stable than filled earth, and therefore, cut slopes are generally permitted to be steeper than fill slopes.

Filled earth may be used where relatively flat, low areas must be raised to make them usable. However, filled earth has several drawbacks: it is not always suitable to support new construction, it often requires expensive compaction, and it is frequently subject to slippage and erosion. If one has a choice, the cut and fill method of grading is generally preferred because it is the least expensive and the most flexible.

The location of the level building area and, to some extent, even the area's finish grade is somewhat arbitrary. There are often several solutions to the design of a site and one may have to proceed by trial and error. Therefore, if grading does not result in a perfect solution the first time (precipitous slopes, destroyed trees, excessive grading, etc.), one should remain flexible and try again.

Circulation Paths

In general, circulation paths should be as level as possible, simply because it takes less effort to move horizontally than up a slope. Natural grade, however, is almost never level; therefore, circulation invariably requires sloped paths as well as level ones. In selecting a route for circulation, one can move parallel to the contours, which is level but requires extensive grading because of its length, or perpendicular to the contours, which results in the steepest path but requires less grading. Normally, circulation paths are graded somewhere between perpendicular and parallel to the contours.

Before laying out a path, one must determine its required width and the maximum allowable steepness. In addition, the slope of the path should be as uniform as possible. This means that contour lines along the path's route will be more or less evenly spaced. It is usually best to

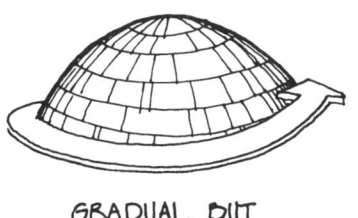
GRADUAL BUT LONGEST ROUTE

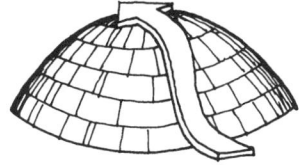
QUICKEST AND SHORTEST ROUTE

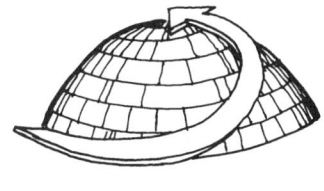
MOST NORMAL ROUTE

CIRCULATION PATHS

locate paths in valleys or along stream beds and avoid placing them on steep slopes. Highways commonly run long distances around a mountain, rather than going directly over or through it and creating extensive grading or tunneling problems.

Grading circulation paths is accomplished the same way as creating level areas, that is, one can cut, fill, or use a combination of cut and fill.

Cutting a path in the slope requires first locating the path to scale on the topographic map. Next, find the intersections of the contours with the low side of the path and extend each contour perpendicularly across the path. When you reach the other side, run next to the path until each contour connects back to its corresponding existing contour line. This will create a graded bank on the uphill side of the path, the steepness of which can be modified through contour manipulation.

Filling to create a path works in the reverse way. After laying out the proposed path in plan, locate the intersections of the contours with the path's high side. As before, cross the path perpendicularly with each contour, run them next to the low side of the path, and reconnect them to their corresponding existing contours.

Using a combination of cut and fill is the most common way to grade a path on sloping terrain.

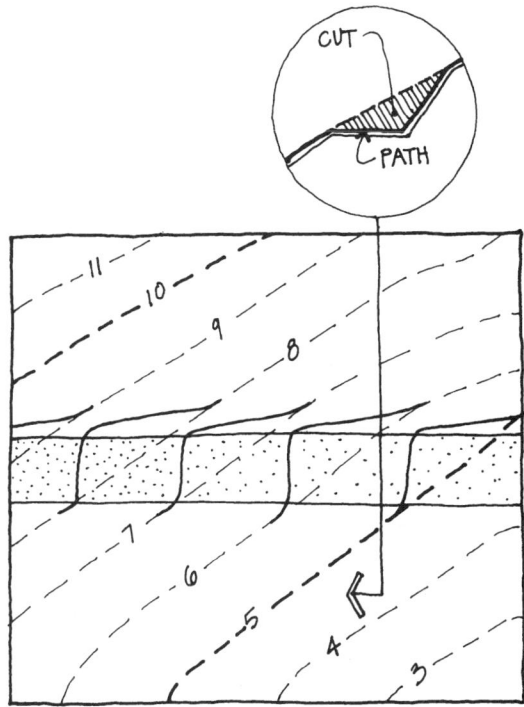

CUTTING A PATH

The procedure here is to lay out the path, select a contour in the middle, establish a perpendicular contour that crosses it at its midpoint, and then reconnect the contours on both sides. By so doing, you will create a cut bank above the path and a filled bank below it.

If the natural grade is not excessively steep, paths may be designed to run perpendicular to the contours, that is, straight up the slope. After the path is located on the topographic plan, the path's

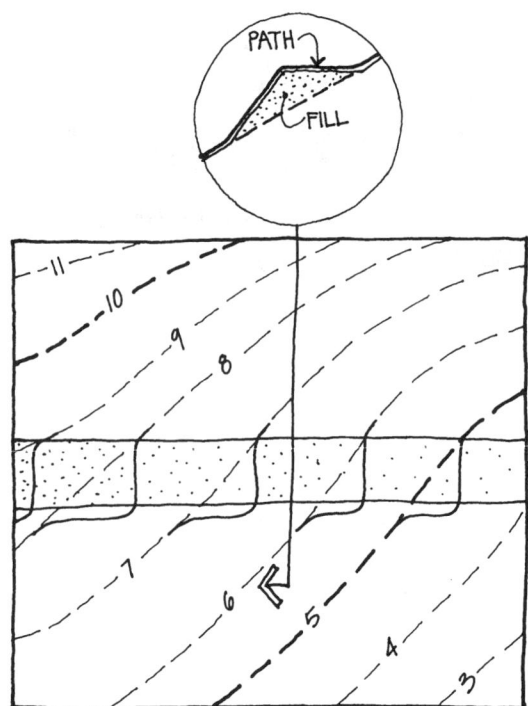

FILLING A PATH

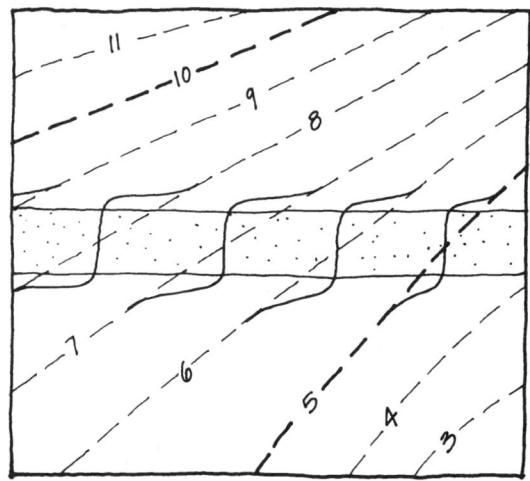

CUTTING & FILLING A PATH

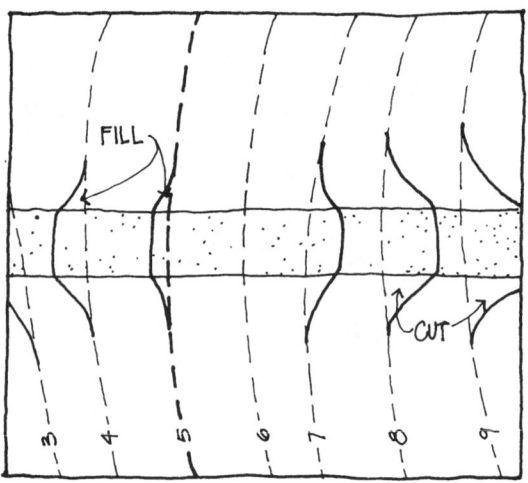

PERPENDICULAR PATH

of the path. The result of grading a path perpendicular to the contours may be either cut banks or filled banks on both sides of the path.

The Grading Process

The grading process begins with the development of a grading plan. This plan shows the site boundaries, existing topography and site features, and the proposed modifications, including new structures. Grading plans use contours, notes, and a variety of special symbols to describe what must be done to reshape the land into the desired forms.

The principal aim of all grading work is to make the land appropriate for its purpose and to preserve a stable system. The finished grade should have positive drainage, stable slopes, balanced cut and fill, and pleasant and harmonious visual forms.

Grading operations begin by removing the topsoil, which is stored on the site and later reused over the modified ground forms. Grade stakes are then placed at intervals in the subsoil to indicate the required new levels. Grade stakes are located at all critical points, such as peaks, valleys, roads, walls, and other points of grade change. Grading machines then cut or fill the

steepness must be established by evenly spacing contours perpendicularly along the path's route. It is best to begin in the middle of the path's length and allow that contour to run right through undisturbed. There may be times, however, when this is not possible, and you may have to begin with an established elevation at one or both ends

earth to the staked levels and shape it into the desired configuration. Among the machines employed for this purpose are bulldozers, scrapers, graders, power shovels, rollers, and scarifiers. With the extensive range of machinery available today it is rarely necessary to do any expensive hand shoveling.

The following summarizes the most important general rules of grading:

1. Do not extend grading beyond the property lines.
2. Strip and save all topsoil prior to grading.
3. Avoid the destruction of valuable existing vegetation.
4. Attempt to balance all cut and fill.
5. Avoid flat grades that create drainage pockets.
6. Avoid erosion by grading slopes within their natural angle of repose.
7. Be certain finish grades enable water to flow away from all structures.
8. Avoid grading solutions that rely on expensive retaining walls, steps, or other construction.

There are several serious hazards associated with grading that may cause irreversible damage. Some of these are as follows:

1. *Loss of topsoil.* It takes hundreds of years of natural decomposition of organic material to produce a thin layer of topsoil. Since it is absolutely essential to plant life, topsoil must be retained.
2. *Loss of vegetation.* Vegetation replenishes the oxygen, moderates the climate, and helps control erosion. All mature plants, therefore, must be preserved.
3. *Altered drainage patterns.* Modified runoff patterns can cause erosion and contamination of downstream waterways. New drainage patterns must be carefully planned.
4. *Unstable earth.* Grading earth that is unstable can produce slides, slippage, and cave-ins. Work done at one location can affect other sites, even if relatively distant.
5. *Aesthetic damage.* Grading that alters existing site qualities may destroy the uniqueness of an entire area. Designers should remain sensitive to existing conditions.
6. *Unique conditions.* Some areas are excessively steep or contain extensive rock outcroppings; others provide a natural habitat for wildlife. Such areas, if possible, should be left in their natural state and not graded or developed in any way.

Although grading is sometimes considered a mechanical process, the site designer who creates a grading plan is actually modeling earth forms in much the same way as an artist sculptures clay. Sensitive grading represents the best transition between the existing land and the proposed requirements of the project. It should be functional, economical, and attractive, and do a minimum amount of ecological damage.

DRAINAGE

Storm Drainage Systems

Drainage of land refers to the methods used to collect, conduct, and dispose of unwanted rain water. During a rainstorm, water is absorbed by the soil in varying amounts and at varying speeds of percolation, depending on the soil's porosity and the volume of soil above the ground water table (see page 43). If rain falls faster than the soil is able to absorb it, puddles form and surface water begins to flow downhill. The uncontrolled flow of this surface runoff may

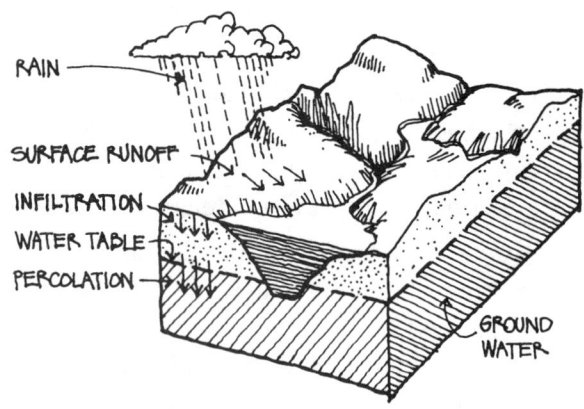

NATURAL DRAINAGE

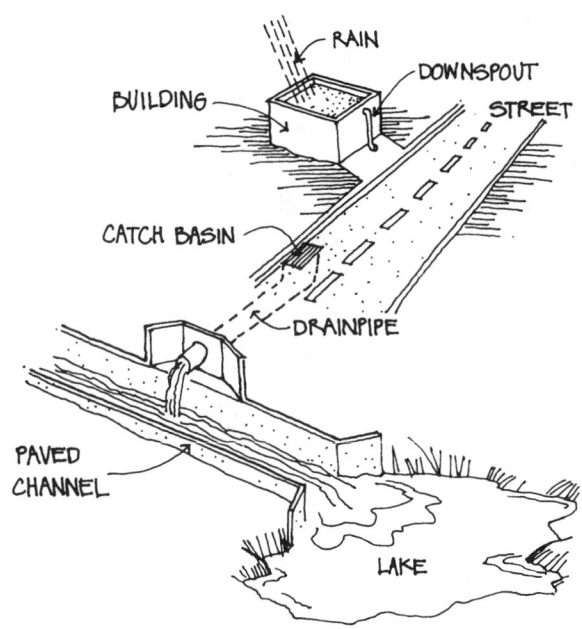

STORM DRAINAGE SYSTEM

result in erosion, flooding, or damage to landscape and structures.

Storm drainage systems have as their objective the removal of excess rain water. They are a substitute for natural surface drainage and are generally necessary only where development is dense enough to cause excess runoff following a rain. Low-density developments often rely on natural open land, but higher densities imply more paved areas, which cause greater runoff.

A typical drainage system begins with the roof water from an individual building. This water flows to roof drains and downspouts that eventually conduct it to the street, either directly or over paved areas or lawns. Once in the street, the water flows downhill until it reaches a catch basin, where it continues in an underground drain line. The drain line may lead to a concrete channel that ultimately discharges the original roof water, plus all other runoff, into a lake or other body of water.

Storm drainage systems are designed to achieve the following:

1. Reduce flooding and damage by eliminating excess rain water.
2. Reduce erosion by controlling the rate and volume of water flow.
3. Eliminate standing water which may lead to pollution and insects.
4. Enhance plant growth by reducing soil saturation.
5. Improve the load-bearing capacity of soils.
6. Cleanse the storm water by installing native plants in drainage ways to act as filters to remove particulates from the flowing surface water.

Underground storm drainage systems are expensive. Therefore, every effort should be made to employ surface systems, such as gutters, ditches, channels, short culverts and, especially, spacious planting areas.

Drainage Needs

The demand for drainage is determined by the following factors:

1. *Topography.* Steep areas drain quickly, often too fast to allow for percolation. Therefore, drainage benches or channels

should be provided above and below steep banks to collect runoff water.

2. *Type of soil.* Soil type determines the amount and speed of water absorption, which in turn affects the runoff quantity.
3. *Vegetation.* Thick ground cover slows down the rate of runoff, reduces erosion, and reduces the need for elaborate drainage systems.
4. *Rainfall data.* Local rainfall data is necessary to calculate the frequency and intensity of rain water to be drained.
5. *Land use.* In rural areas, one can generally allow water to disperse over the landscape. In urban areas, on the other hand, surface runoff occurs for short distances only and then must be directed to sub-surface drainage devices.
6. *Size of area.* The area to be drained generally refers to those areas with limited percolation, such as roofs, roads, driveways, etc. Fields and open spaces can usually remain wetter than parking areas and paved walks.

Accurate runoff flows can be computed by use of the Rational Formula, which is $Q = CIA$, where Q is the quantity of runoff in cubic feet per second, C is the coefficient of runoff, I is the rainfall intensity in inches per hour, and A is the drainage area in acres. The coefficient C is the fraction of total rainfall that runs off a surface. Roofs and pavements have a value of about 0.9, while lawns and other planted areas have a value around 0.2.

Surface Drainage

Surface drainage involves the removal of runoff water by means of surface devices only. It is generally preferred over subsurface systems, because it is less expensive, it allows some water to percolate into the ground, and the danger of clogged pipes is eliminated. The following are some general rules of surface drainage:

1. Water flows as a result of gravity; therefore, all surfaces must be sloped for drainage.
2. Water always flows perpendicular to the contours.
3. Good drainage requires a continuous flow. Slow-moving water may create bogs, while water moving too fast will cause erosion.
4. Water should always be drained away from structures.
5. Large amounts of water should never be drained across circulation paths.
6. Desirable slopes for surface drainage are as follows:

    Open land - 1/2 percent minimum

    Streets - 1/2 percent minimum

    Planted areas - 1 percent minimum to 25 percent maximum

    Large paved areas - 1 percent minimum

    Land adjacent to buildings - 2 percent minimum

    Drainage swales - 2 percent minimum to 10 percent maximum

    Planted banks - up to 50 percent maximum

Following are the most common methods of surface drainage:

1. *Swale.* Sloping areas can be drained by creating swales, which are graded flow paths similar to valleys. Swales are graded around structures with finish contours always pointing uphill and flow paths shown perpendicular to the revised contours.
2. *Sloping plane.* This is the simplest, cheapest, and, consequently, the most common way to drain a relatively level area. The area tilts in one direction, so that the water drains to the low side. Adjacent structures are always located at the high side.

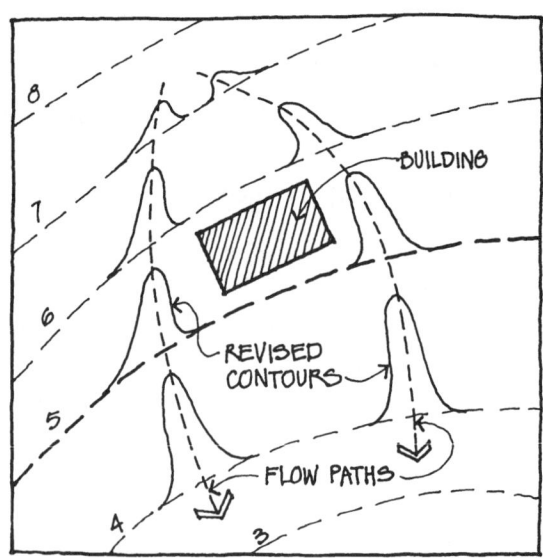

CREATING SWALES

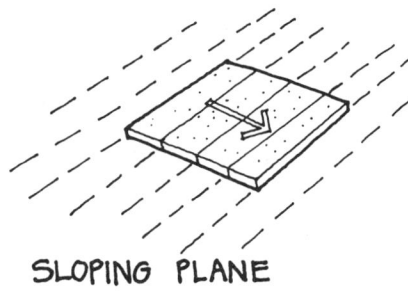

SLOPING PLANE

3. *Warped plane.* The high side is level, similar to the sloping plane. The contours, however, are fan-shaped, so that the entire area drains to one low corner.

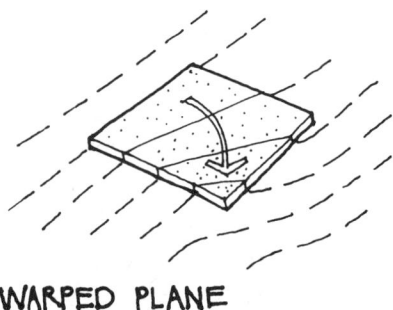

WARPED PLANE

4. *Gutter.* Gutters are formed by two sloping planes that create a valley. The planes are slightly warped so that water can run down the valley to a collection point. When adjacent to a structure, the top edge of one sloping plane will be level.

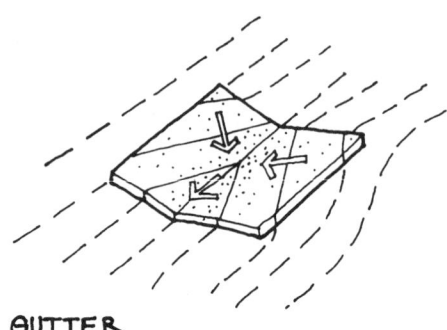

GUTTER

5. *Central inlet.* Large flat areas, especially where enclosed (courtyards, patios, etc.) employ a central drain toward which all surfaces slope. The disadvantage of this arrangement is that it requires a catch basin and sub-surface piping to dispose of the water.

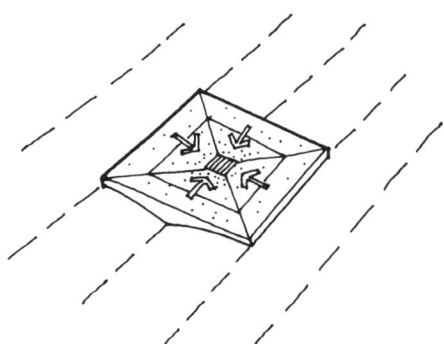

CENTRAL INLET

Sub-Surface Drainage

Sub-surface drainage refers to the collection, conduction, and disposal of water below ground level. Runoff water flows not only on the surface, but also below the ground surface. Collection of sub-surface water utilizes gravel-filled ditches and perforated drain pipe, or drain pipe laid with

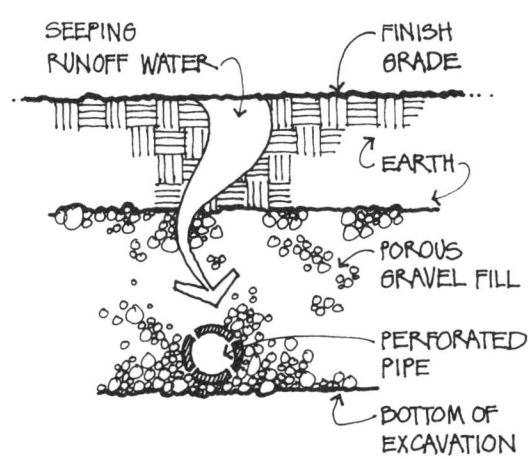

SUB-SURFACE DRAINAGE

directly to underground pipes. It has a metal grate to prevent debris from entering and clogging pipes.

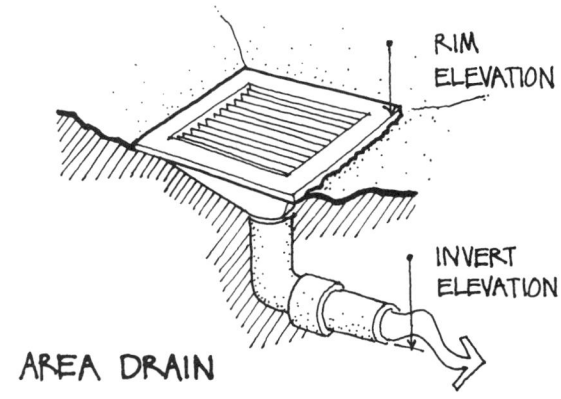

AREA DRAIN

open joints. Runoff water that seeps into the earth flows vertically through the gravel until it reaches the openings of the drain pipe. This pipe collects the free-flowing water and carries it away in the direction of the sloping pipe.

Closed sub-surface systems consist of various fabricated collectors together with sections of closed pipe, which are used to carry water below grade from collection points to disposal areas. Such systems are useful in level areas, since drain lines can be buried progressively deeper, assuring an adequate slope for drainage.

Among the common fabricated collectors are the following:

Area drain. This device collects water from the low point of a limited area and conducts it

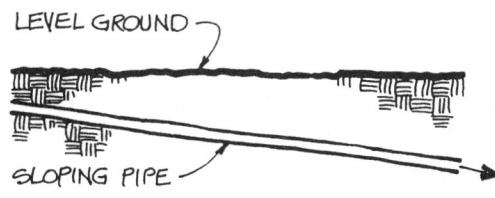

DRAINAGE INDEPENDENT OF GROUND SLOPE

Catch basin. This is similar to an area drain, except deeper and generally larger in order to catch and retain sediment which may clog the system.

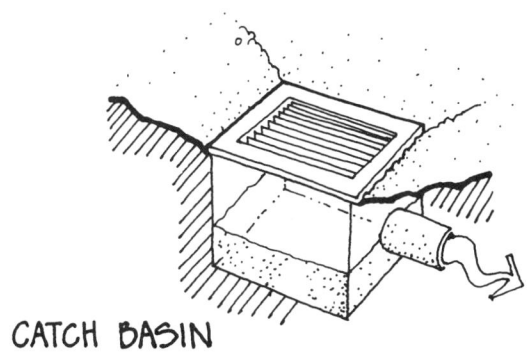

CATCH BASIN

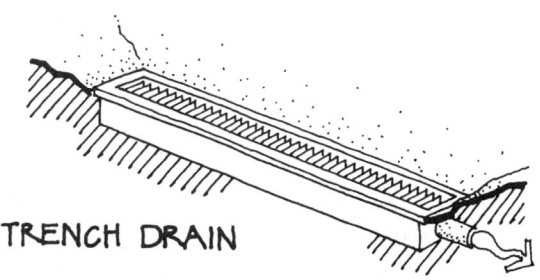

TRENCH DRAIN

Trench drain. This device is used to collect water along a wide strip before conducting it to underground pipes. It is suitable at the entrance to an underground garage, for example, where it collects the runoff water flowing down a sloping driveway.

Drainage pipes are manufactured from clay, concrete, plastic, or composition materials. They are rarely less than four inches in diameter and require a minimum slope of one percent to assure proper flow. On drawings, the depth of a drain pipe is indicated by the invert elevation, which is the elevation of the bottom of the pipe leading out of the drainage collecting device. The top of the device is indicated by the rim elevation.

Pipes which run underground beneath roads, driveways, or paths are referred to as *culverts.* They vary in size from six inches to several feet. Culverts should be straight, cross the road at right angles, and be sufficiently strong to resist moving traffic loads.

CULVERT

Following are some sub-surface drainage general rules:

1. Always determine a site's disposal points before designing the drainage system.
2. Sub-surface drain lines should travel in straight lines; changes of direction occur only at catch basins.
3. Avoid running drain lines beneath or through building foundation walls, retaining walls, or other construction.
4. Drain lines should follow the natural site slope as much as possible to minimize the depth of trenches.
5. Drainage systems often require several lines, and a branching system is frequently the most efficient solution.

DESIGN METHODS

Gradients

Gradient refers to the slope of land, which may be expressed as a ratio or a percentage, or occasionally in degrees. It is useful to know the gradient of a landform for a number of reasons: this information is used in slope analysis, maximum slopes are specified this way in building codes, and different soils have varying angles of repose, that is, the steepness beyond which soil will slide.

Gradient ratios express slope as the ratio between a horizontal distance and the vertical elevation change within that distance. By convention, the first figure of the ratio represents the horizontal distance, while the second figure (reduced to a factor of 1) represents the vertical elevation distance. For example, if land rises 20 feet in a horizontal distance of 40 feet, the slope is 40:20, which is expressed as 2:1.

As another example, if the horizontal distance between the 6 foot and 10 foot contours on a topographic map scales 12 feet, the slope is 12:4 (10 minus 6), which is expressed as 3:1.

Gradient ratios are generally used for slopes on small-scale projects, and also to express design standards, such as:

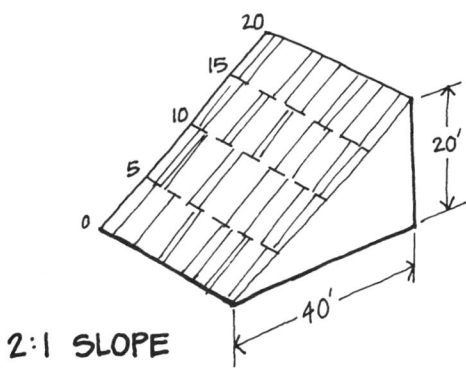

2:1 SLOPE

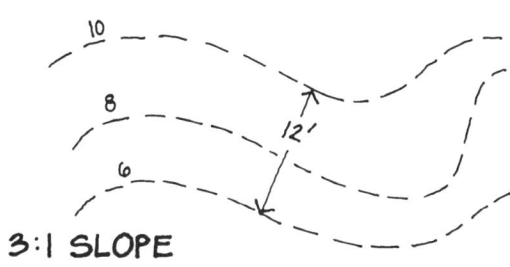

3:1 SLOPE

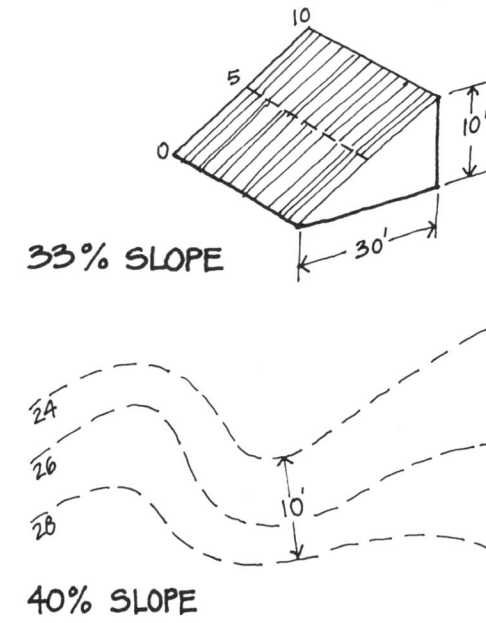

33% SLOPE

40% SLOPE

2:1 - maximum allowable ground slope with little danger of erosion

3:1 - maximum slope for most planted areas

4:1 - maximum slope maintainable with a lawn mower

Gradient percentages express slope as the gradient (G) obtained by dividing the vertical elevation change (V) by the horizontal distance (H) in which the change takes place, expressed as a percentage. Therefore, G = V/H, where

   G is the gradient in percentage

   V is the vertical rise

   H is the horizontal run

For example, if land rises 10 feet in a distance of 30 feet, the slope is 10/30 = .33, which is expressed as 33 percent.

Likewise, when scaled from a topographic map, if the horizontal distance between the 24 and 28 foot contours is 10 feet, the slope is 4/10 = .40, which is expressed as 40 percent.

The percentage method is more widely used than the ratio method, and it is also used to express design standards, such as:

0–1%   Flat slope, poorly drained, generally undesirable for development.

1–5%   Considered ideal for most development.

5–10% Suitable for most development, maximum for walks.

10–15% Considered too steep for many land uses without grading.

15%+   Too steep without substantial development efforts and costs.

The percentage method may be used to calculate the vertical rise or horizontal run when one or the other of these values and the gradient are known. For example, if the gradient of a road is 6 percent and the elevation at point A is 21.7, then the horizontal distance X from point A to contour 22 is calculated as follows:

X = H = V/G or H = (22–21.7)/0.06 = 0.3/0.06 = 5 feet

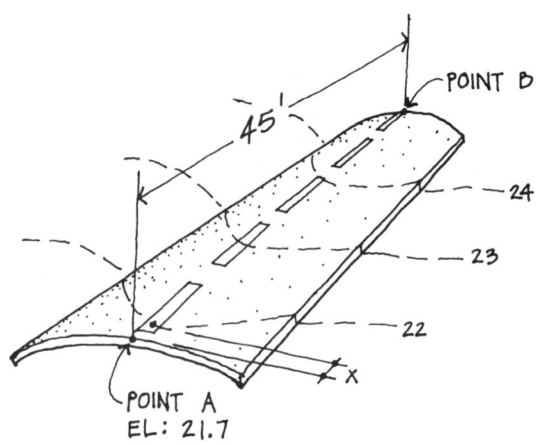

ROAD WITH 6% GRADIENT

Similarly we can calculate the elevation at point B if the distance between points A and B is 45 feet, as follows:

V = GH or V = 0.06(45) + 21.7 = 2.7 + 21.7 = 24.4

Percent of slope must never be confused with angle of slope. For example, a 45-degree slope, which is very steep, has a slope ratio of 1:1 and a slope percentage of 100 percent.

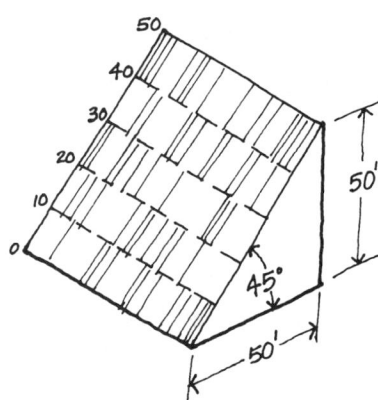

45° - 100% - 1:1 SLOPE

Calculating Cut and Fill

It is sometimes necessary to calculate the amount of cut and fill caused by modifying the existing topography. It is at this point that designers discover that moving a contour line half an inch on a plan may require moving a ton of earth 50 feet. Grading calculations are necessary to verify the balance between cut and fill, as well as to determine the cost of grading.

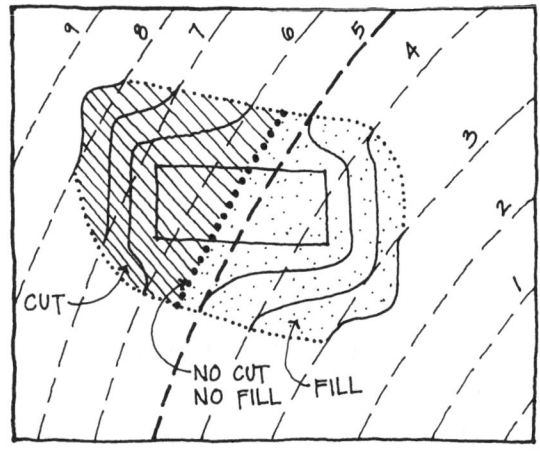

CUT AND FILL

Cut and fill quantities are expressed in volume, generally cubic yards of earth, and they are calculated as width × length × depth. The problem with calculating volumes of earth, however, is that contour shapes are almost always irregular. Therefore, quick calculations will produce only approximate results.

To determine the volume of cut or fill, it is necessary to calculate the area of a particular contour's horizontal surface (width × length) and then multiply it by the contour interval (depth). This can be done in two different ways. First, the irregular area can be divided by equally spaced parallel lines drawn a convenient distance (d) apart. Starting at one end, you measure the various lengths of these lines (L_1, L_2, etc.) and multiply the average length of two adjacent lines by the width d. The sum of all these areas

$$\left(\frac{L_1 + L_2}{2}\right) \times d + \left(\frac{L_2 + L_3}{2}\right) \times d +$$

etc. will be the total area of the irregular form. This area is then multiplied by the contour interval to obtain the volume.

The second way to compute the area is to overlay the irregular form with transparent graph paper, preferably with the grid paper at the same scale as the drawing. In that case, since each square represents one square foot, you simply count the number of squares within the irregular form to obtain the area. The area is then multi-

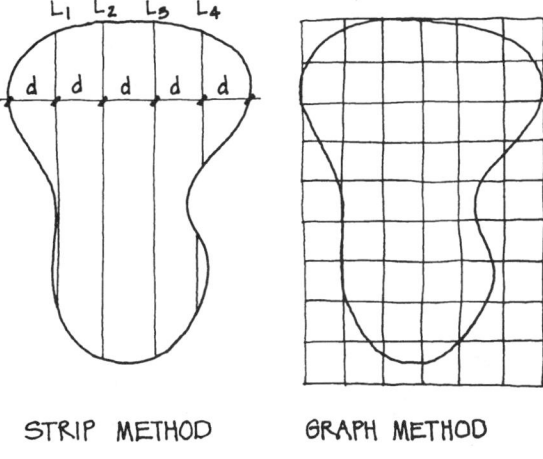

STRIP METHOD   GRAPH METHOD
AREA OF IRREGULAR SHAPES

plied, as before, by the contour interval to obtain the volume.

After each contour volume has been calculated, you add up the cut and fill volumes to derive the total cubic footage of earth for each. These totals are then converted to cubic yards by dividing by 27, which is the number of cubic feet in a cubic yard. The total of these two numbers represents the total amount of earth to be moved; their difference represents the excess of cut or filled material.

Retaining Walls

One common purpose of grading is to create flat, usable areas on which to place something. In some cases, where earth banks are as steep as permissible and the amount of available land is limited, one may be forced to construct a retaining wall. A retaining wall creates a level area by cutting vertically through a bank and eliminating the slope necessary to accommodate a soil's angle of repose. Retaining walls are generally constructed of masonry or concrete, although other materials, such as rock, timber, or steel may also be used. A retaining wall is generally an expensive way to create a level area, but where land is costly, the expense may be worthwhile.

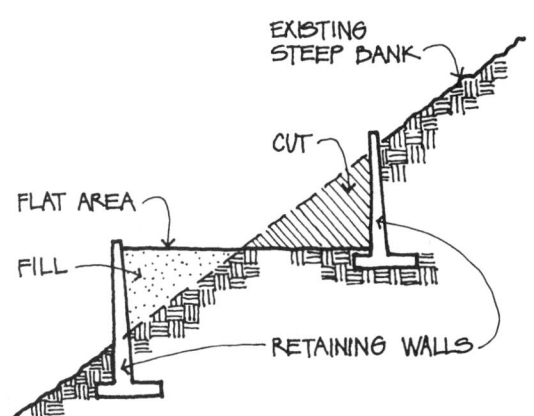

CREATING A FLAT AREA

When shown in plan, the contours that are cut by a retaining wall appear to stop at one end and emerge at the other. In reality, however, the contours travel along the wall and overlap one another to form a vertical plane of earth that is retained by the wall itself. Spot elevations are usually used to indicate the elevation at the top of wall (T.W.) and bottom of wall (B.W.).

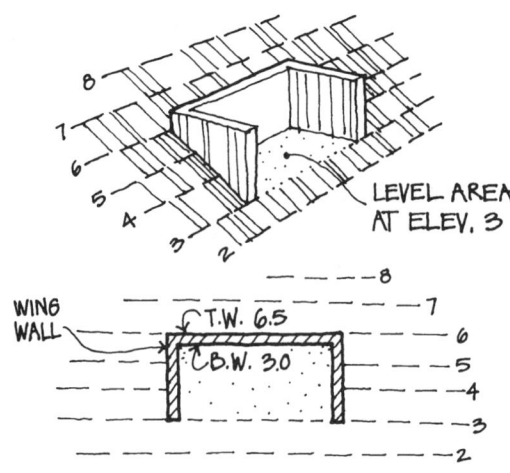

RETAINING WALL

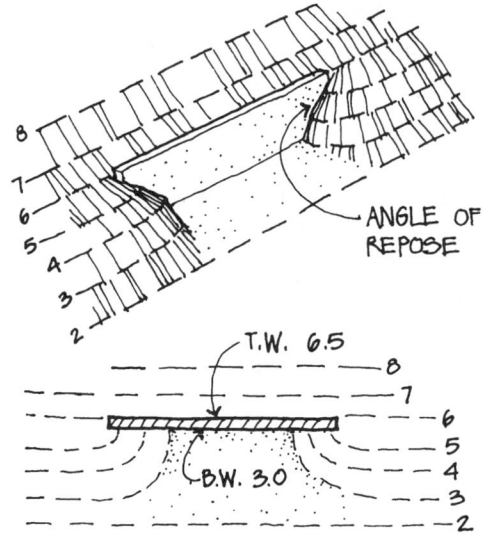

RETAINING WALL

Wing walls are frequently used to accommodate the sloping bank as it returns to the original grade at the ends of the wall. Without wing walls, the bank would wrap around the wall and slope down to the original grade. In that case, the angle made by the soil along the face of the wall should not exceed the soil's angle of repose.

Roads

For safety and efficiency, roads are constructed as level as possible; nevertheless, as with pedestrian paths, roads must be slightly sloped to insure adequate drainage. Road drainage is accomplished in the following ways:

Sloped. The road is graded level across the width but sloped over the length. This causes water to flow in a sheet down the length, where at some point it is collected and conducted elsewhere. Contours are shown crossing the road perpendicularly and break in the downhill direction on each side where the curb is raised.

Pitched. The road is pitched toward one side allowing water to flow along one gutter. This method is advantageous where the road parallels the contours, since some of the grade may be taken up by tilting the road in the same direction

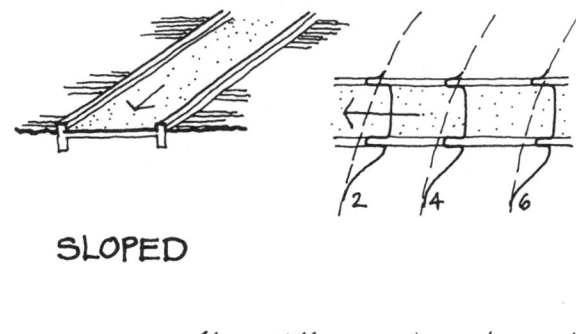

SLOPED

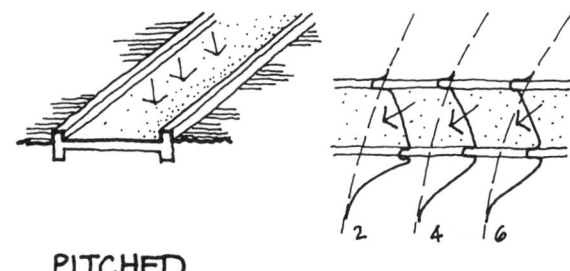

PITCHED

as the land slopes. Contours cross the road at an angle, and the water flows perpendicular to the contours.

Central Gutter. The road is depressed in the center, which allows the water to run to the middle and flow to some collection area. Contours are drawn similar to a valley, that is, with the contours pointing in the uphill direction.

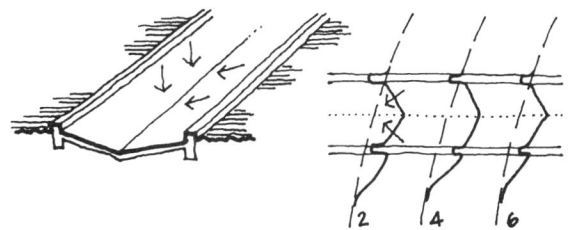

CENTRAL GUTTER

Central Crown. The road is a convex curve with water pitched toward both sides. This is the most common method used, and a six-inch high crown is typical in a 40-foot street width.

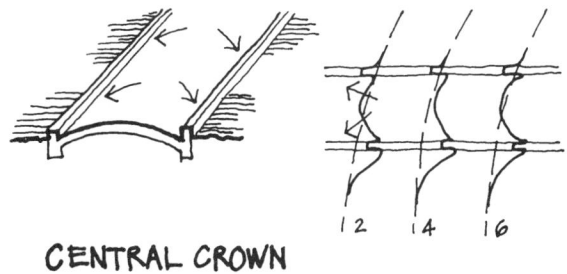

CENTRAL CROWN

Contours are curved and point downhill, similar to the indication for a ridge.

Site designers often have difficulty visualizing contours along a road. Shown below is a contour running across a road with a crown and a raised curb. The contour line curves in the downhill direction, runs level along the curb face, and reappears on top of the curb a little further downhill. Remember, every point along the contour line is at the same elevation.

Another example shows a road with a crown and the roadbed elevated above the adjacent grade. The contour line again curves downhill, but reappears on grade a short distance uphill.

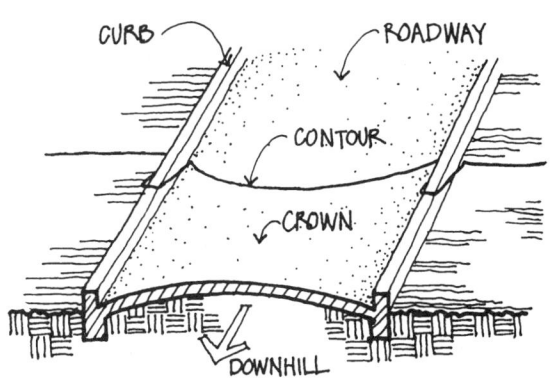

ROAD WITH RAISED CURB

Other such configurations can be analyzed and envisioned similarly.

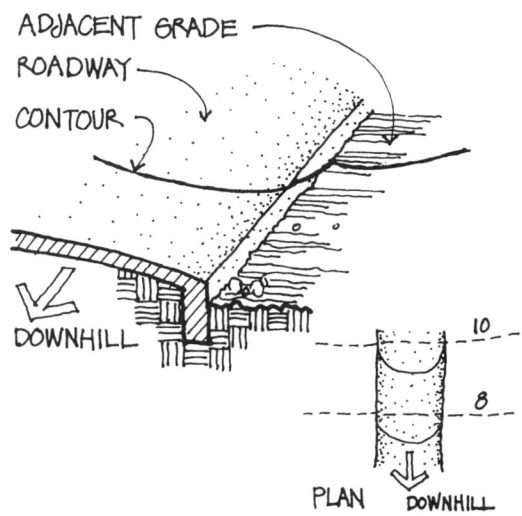

ELEVATED ROAD

Parking Lots

One of the principal objectives of grading is to provide satisfactory surface drainage so that water will be collected and conducted away from the site. For a relatively flat area, such as a parking lot, there are several acceptable ways to do this. First of all, large paved areas should have a minimum grade of 1 percent and a maximum grade of 5 percent. Less than 1 percent will result in occasional ponding, while a slope

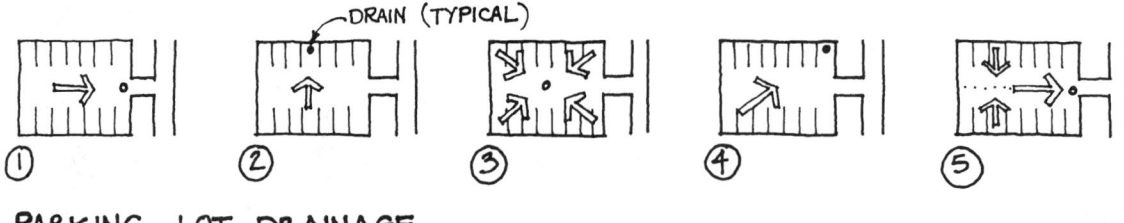

PARKING LOT DRAINAGE

exceeding 5 percent may allow surface water to flow fast enough to cause erosion or damage.

Parking lots are drained in the following ways:

1. Tipping the entire lot and allowing the water to flow (sheet) along the entire lot length.
2. Tilting the lot and draining the water to one side.
3. Sloping the lot from all corners and collecting the water at the center in a funnel fashion.
4. Warping the lot and collecting the water in one corner.
5. Depressing the lot at the center and draining the water toward one end along the center line.
6. Tilt each double row towards a center grassy area with selected wetland vegetation. This process cleanses the runoff from automobiles before it enters the main storm water collection area.

In no event should large amounts of water be permitted to flow across sidewalks adjacent to the parking areas. Catch basins or trench drains should be strategically placed to collect flowing water and conduct the excess runoff through the storm drainage system.

Building and Topography

It is always easier and more economical to locate a building on a relatively level site than on one that is steeply sloping. On a level site one has flexibility in the arrangement of building forms. As the ground surface becomes steeper, the site becomes more restrictive and development costs increase. The way a building relates to a sloping site depends largely on the steepness of the slope. Generally, the building footprint should run with its long axis parallel to the contours. This arrangement minimizes the amount of grading required to fit the building to the ground.

The ground directly adjacent to the base of a building should slope away for a short distance so that surface water flows clear of the structure. This may necessitate creating a swale on the

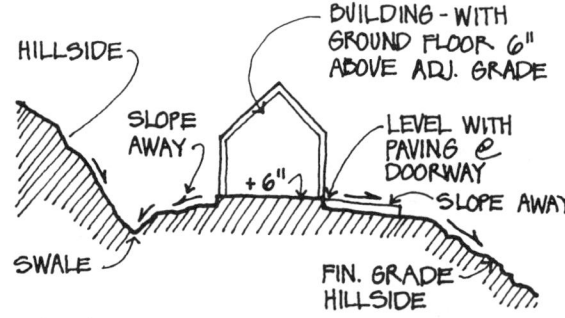

SURFACE WATER CONSIDERATIONS

uphill side of the building in order to catch the flowing water and divert it around the building. In addition, the ground floor should always be set at least six inches above the adjacent finish grade to reduce the possibility of water inflow. The situation at access points, of course, is an exception; the finish grade at doorways should be at approximately the same level as the ground floor to allow easy access for the handicapped.

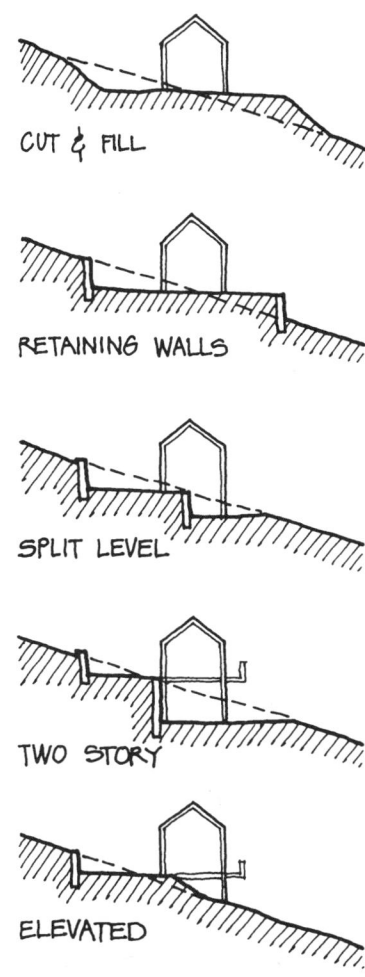

BUILDING ON A SLOPE

There are several ways to adapt a building to a slope, such as:

1. Create a level area by cutting and filling.
2. Create a level area by using retaining walls.
3. Create two level areas and use a split level design.
4. Create two level areas and use a two-story design over one of them.
5. Create a level area, project the building beyond this area, and elevate it above the ground.

In order to achieve a totally successful design solution, buildings must relate to their site both functionally and aesthetically. Designers, therefore, must use their technical knowledge, as well as professional judgment, in dealing with the problems and opportunities created by topography.

SITE ANALYSIS

General Purpose

Site analysis is the process of investigating basic data that relates to a particular site, such as survey information, topographic data, geological information, zoning ordinances, existing character, microclimate, development patterns, social patterns, etc. The purpose of site analysis is to determine whether a parcel of land is suitable for a specific proposed use. It would be undesirable, for example, to situate a school adjacent to a major freeway. Similarly, a roadside cafe should not be located out of sight of the road, nor should a meat packing plant be placed upwind of dwelling units. All of these are examples of inappropriate uses for a given site, or perhaps inappropriate sites for a given use. In theory, almost any site will support almost any use; however, the ideal situation is one that most fully satisfies the project criteria, while requiring the least modification or compromise.

Every site is as unique as an individual person, and even as identical twins have distinct personalities, so too, adjacent parcels of land possess distinguishable characteristics. Part of this uniqueness is reflected in a site's equilibrium. The flow of surface water creates a discrete drainage pattern, plant and animal life constitute an ecological system, and human use conforms to a workable social structure. Site factors such as these are interrelated, and at any given moment they are in balance, even if they are in the process of change. The recognition of a site's character reveals the practical limits imposed on a planner, as well as the potential damage that may be inflicted by development.

All development implies change, and occasionally this change produces undesirable effects. Excavation, for example, may alter drainage patterns, grading may cause erosion, and the construction of new facilities may destroy plants, pollute the air, or create traffic congestion. Environmental changes, therefore, are an inevitable result of the development process.

Relevant Data

The relevant site data which must be gathered and analyzed comprise those factors that determine the suitability of a site for its proposed use. Although no single set of factors applies to every situation, the following list includes data that is relevant in most circumstances. Some of these factors are described in greater detail in the previous lessons.

Climate

Every site is affected by regional climate patterns, as well as the microclimate that applies to a small area. Climate is related to topography, slope orientation, vegetation, and the presence of water, and it is important because it bears directly on human comfort. Climatological data may be obtained from the National Weather Service, through talks with local inhabitants, and by personal observation of weathered structures and existing plant material. The following features may be analyzed:

1. Temperature averages and extremes
2. Precipitation averages and extremes
3. Snowfall averages and extremes
4. Wind intensity and directions
5. Humidity patterns
6. Solar angles
7. Days of sunlight
8. Frost data

Topography

Topography is the form of a site's surface features, and it is a factor that strongly influences land development. The gradient of roads, disposition of structures, and visual aspects of a site are all influenced by the character of the landform. Topographic data is available from the U.S. Geological Survey aerial photographs, or on-site surveys, and the features that may be analyzed are:

1. Elevations
2. Slope amount and direction
3. Unique landforms
4. Natural drainage patterns

Soils

Knowledge of the soil conditions on a site is important to determine the soil's capacity to support buildings and roads, as well as its ability to sustain plant material. Soils data is obtained from the U.S. Department of Agriculture Soil Conservation Service, test borings, visual inspection, and the experience of neighboring developers. The following features may be analyzed:

1. Soil types
2. Moisture content
3. Depth of organic topsoil
4. Depth to water table
5. Depth to bedrock
6. Drainage characteristics
7. Susceptibility to compaction
8. Soil fertility
9. Rock outcroppings

Hydrology

Hydrology refers to the occurrence, movement, and quality of water on a site. Surface water and drainage patterns affect vegetation, climate, and potential development, and this data is available from the U.S. Geological Survey, local hydrological studies, and on-site inspections. Hydrological considerations include:

1. The form of surface water (streams, lakes, etc.)
2. Drainage patterns
3. Runoff rates
4. Subsurface water characteristics
5. Aquifer (water-bearing) zones

Vegetation

Plant types and patterns represent a major site resource, and they contribute significantly to the unique character of an area. Native landscaping is closely related to climate, hydrology, and topography, and it often determines the form of development.

Data on vegetation is available from U.S. Geological Survey maps, aerial photos, and on-site observation. The factors that may be analyzed are:

1. Types and extent of vegetation
2. Density of vegetation
3. Heights of vegetation
4. Health of vegetation

Existing Land Use

As a site is developed, man-made features become more important than natural features. Structures, circulation systems, and activity patterns must be considered. Such data is obtained from land use maps, state highway maps, historical preservation societies, and personal inspections, and some of the factors that may be analyzed are:

1. Existing roads and paths
2. Existing utility lines
3. Existing air and rail facilities
4. Type and number of structures
5. Uses of open space
6. Human behavior patterns
7. Historical sites, structures, and trails

Sensory Qualities

The sensory qualities of a site are those intangible elements that affect people through the senses of sight, smell, touch, and hearing. The uniqueness of a site may be its view or its geometry, the smell of wildflowers or of the ocean, the feel of heat or of wind, or the sound of traffic, church bells, or singing birds. The perception of sensory qualities is as important to site analysis as any other relevant factor, and pertinent data of this sort is almost always obtained through first-hand on-site observation. The features to be analyzed may include:

1. Scenic vistas
2. Spatial illusions
3. Quality of light
4. Characteristic smells
5. Characteristic sounds (noises, echoes, etc.)
6. Sensation of natural forces
7. Perception of textures

Natural Hazards

There are several natural elements that are potentially hazardous to certain types of development, and others, such as earthquake faults, that may restrict almost all construction. Information on hazards is generally available from a variety of government agencies, local inhabitants, and sometimes (unfortunately) through personal experience. Analysis may include:

1. Earthquake fault zones
2. Hurricane zones
3. Tornado zones
4. Flood plains
5. Tidal inundation areas
6. Wet zones (peat bogs, quicksand, etc.)
7. Areas of poisonous plants
8. Areas of poisonous snakes or reptiles
9. Areas of annoying insects

The actual site analysis begins when all of the pertinent information is collected. At this point, a base map is prepared showing legal boundaries, contours, roads, buildings, utilities, and other natural or man-made key features. The base map is used as a background on which various overlays are produced, generally one for each area of concern. For example, a soils overlay may classify soils by type and depth, with locations and logs of known test borings.

A visual survey overlay may consist of personal notes and observations regarding scenic views or unsightly features in need of modification or removal. When environmental concerns are explored, the resulting overlay may serve as a checklist for an environmental impact assessment.

A map on which all the overlaid information has been superimposed is known as a site analysis map, an example of which is shown on the following page. This map indicates the degree to which a site is suitable for a proposed function. At this point, the planner may discover that compromises may be necessary. For example, a site that appears optimum for a shopping center, based on population growth studies, topography, suitable soil, costs, etc., may be located too far from freeway access. For the shopping center and freeway access to be closer together, one may be forced to accept a lower quality site. Therefore, when a site is judged to be suitable for a proposed use, it is almost always a matter of striking a balance between what is ideal and what is reasonably possible.

Lesson Four: Site Planning Issues: Assessing the Site

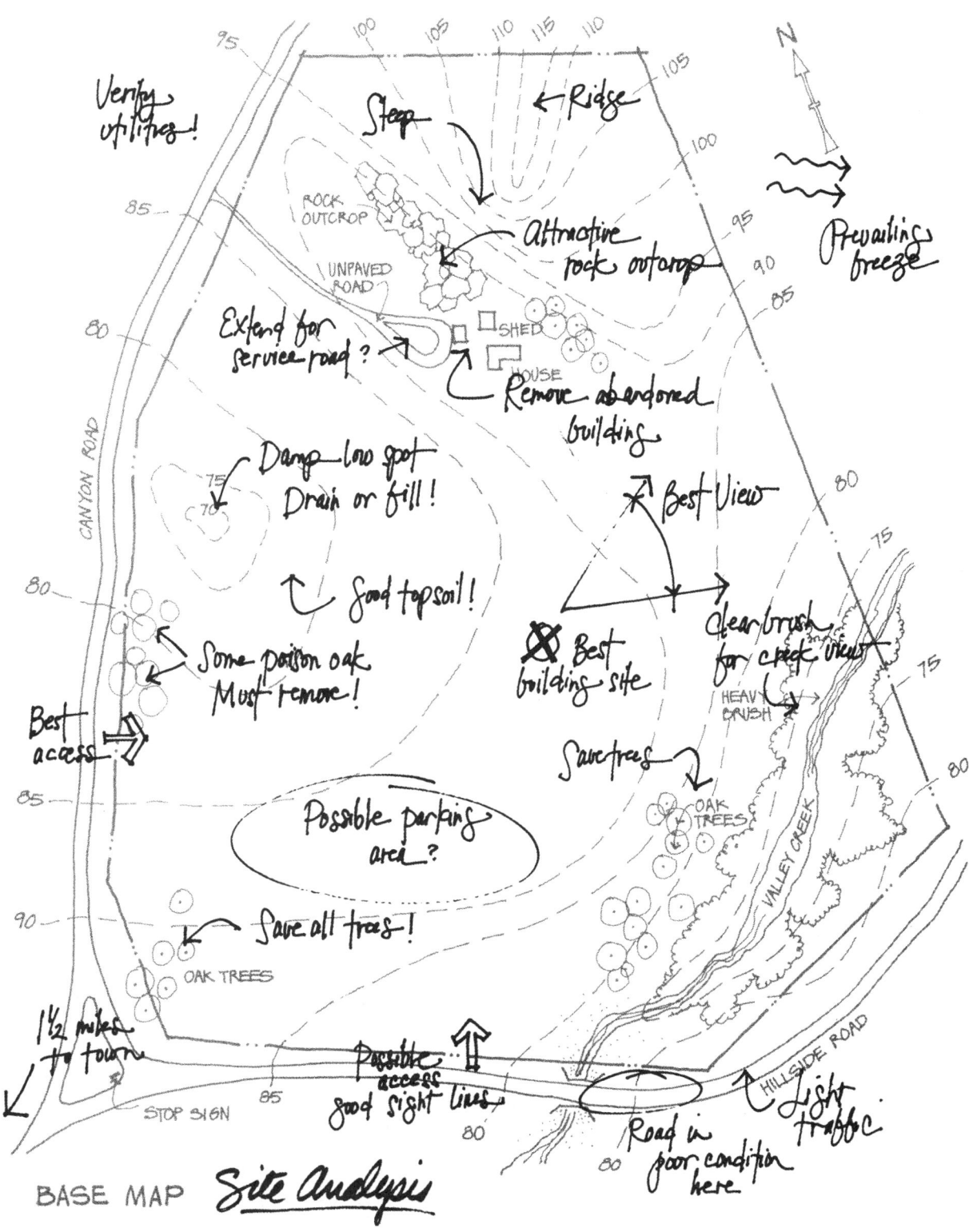

BASE MAP Site Analysis

LESSON FIVE

SITE PLANNING ISSUES: DESIGNING THE SITE

Site Design Process
Circulation Systems
   Introduction
   Design of Vehicular Circulation
   Parking
   Design of Pedestrian Circulation
Zoning
   General
   Zoning Envelope
   Setbacks and Yards
   Height Limitations and Variable Setbacks
   Land Coverage
   Floor Area Ratios
   Off-Street Requirements
Flexible Zoning
   General
   Variances and Conditional Uses
   Rezoning
   Contract Zoning
   Bonus or Incentive Zoning

SITE DESIGN PROCESS

The site design process is an exploration of possible solutions to a specific problem. This exploration involves a number of essential steps, generally performed in sequence, which ultimately leads to a solution of the project's objectives. In the usual case, a client intends to develop a piece of land for some purpose. The designer may be contacted by the client either before or after the site has been selected. Either way, the designer must become familiar with the client's goals, the intended land use, and the parcel of land itself. From that point on, the sequence of activities includes the following steps:

1. **Project Proposal**
   A. Scope of services
   B. Cost of services
   C. Time of performance
2. **Research and Analysis**
   A. Site inventory
   B. Data analysis
   C. Client objectives
   D. Program preparation
3. **Design Phase**
   A. Circulation pattern
   B. Functional pattern
   C. Form composition
   D. Diagrammatic plan
   E. Schematic plan
   F. Preliminary plan
   G. Master plan (design development)

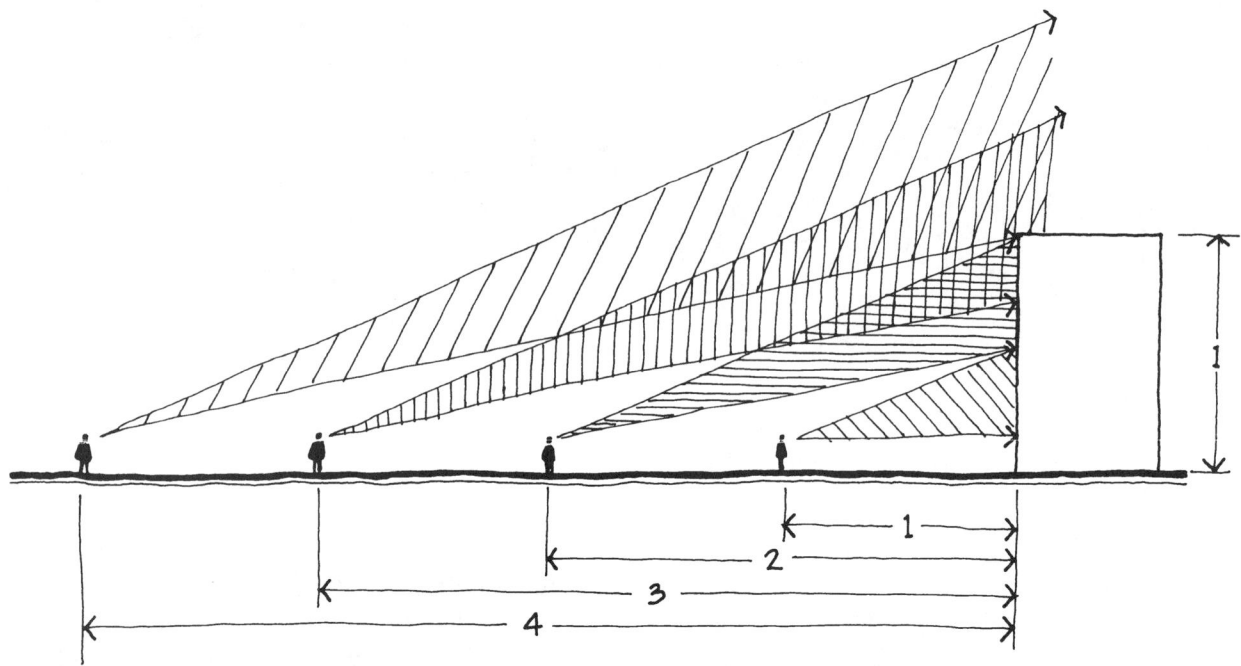

PERCEPTION OF MASS BASED ON VIEWING DISTANCE

4. **Construction Phase**
    A. Technical plan
    B. Grading plan
    C. Landscaping plan
    D. Construction details
    E. Contract documents
5. **Post Construction**
    A. Evaluation
    B. Maintenance

While these various steps occur in sequence, some may overlap or occur simultaneously. Moreover, no step occurs independently of the others. The design process outlined above does not guarantee a beautiful or even functional solution; it is merely a framework of activities that one must perform to achieve an answer to a specific puzzle. The answer may result in a masterpiece, a disaster, or, more than likely, something in between.

Design success relies on a designer's knowledge, inspiration, experience, intuition, talent, ability, and creativity, and these qualities vary with the individual. In this course we have addressed one of the most important factors for successful design: knowledge. We hope, however, that this knowledge will lead to more inspired, creative, and responsible solutions to site problems.

CIRCULATION SYSTEMS

Introduction

Paths of movement, whether of people, automobiles, goods, or services, are linear in nature. They all have a starting point from which they move through and past a sequence of spaces until they arrive at a destination. The shape and form of the path depend on the type of transportation. Pedestrians can turn, pause, stop, and rest at will. An automobile, however, has less free-

Lesson Five: Site Planning Issues: Designing the Site **61**

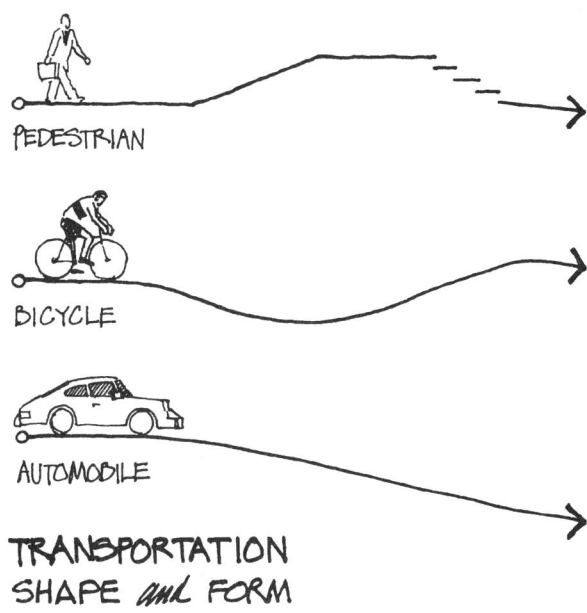

TRANSPORTATION SHAPE and FORM

the designer must provide sufficient space for this. The configuration of roads and walks relates directly to the pattern of buildings and spaces it links. A circulation element that parallels a pattern of buildings reinforces the spatial organization; one that opposes it can act as a visual counterpoint to such a pattern. If one is able to visualize a site's overall layout and configuration of roads and walks, the orientation and spatial arrangement are clearly perceived. Consequently, one of the primary objectives of a circulation system is to lead rather than to confuse.

dom to change direction or come to a sudden halt. Nevertheless, an automobile can negotiate a smoothly contoured path tailored to its physical size.

Pedestrians, on the other hand, although able to tolerate abrupt directional changes, require more space than their physical size and greater freedom to choose their direction of movement. Vehicular traffic can also be more easily controlled than foot traffic, since automobiles do not have as many choices of circulation routes nor the freedom of access that pedestrians have.

When one reaches a branch or an intersection in a walkway or road, one must decide in which direction to proceed. The scale and continuity of each path at an intersection allow one to distinguish between major and minor routes leading to more or less important buildings and spaces. When intersecting roads and walks are similar in size, people must pause, orient themselves, and decide which direction to follow, and

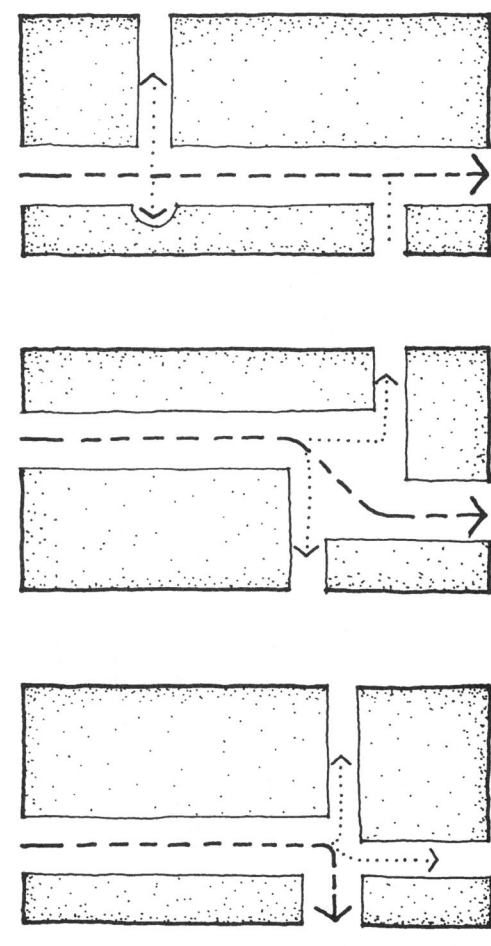

BRANCHES and INTERSECTIONS

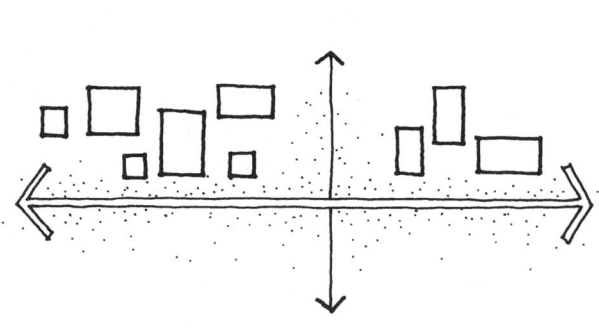
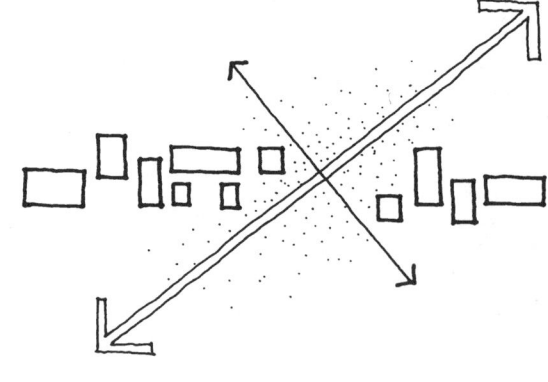

PARALLELS

OPPOSES

Design of Vehicular Circulation

Virtually all developable sites are near or adjacent to public streets. Most urban sites are accessible from streets along one property line, which limits the number of choices for locating points of ingress and egress. Suburban and rural sites usually offer more choices, although they may require the construction of special access roads to connect with existing highways. In every case, the following are primary considerations in locating points of access to a site:

1. Interface points of access with surrounding circulation systems and patterns.
2. Relate access points to existing and future adjacent uses and developments.
3. Avoid potential conflicts between vehicles and pedestrians.
4. Relate access points to on-site parking and service areas.
5. Minimize the environmental impact of points of access on natural features of the site and its surroundings.

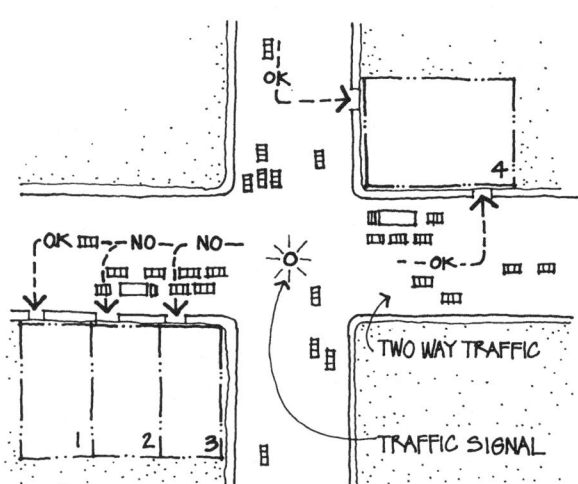
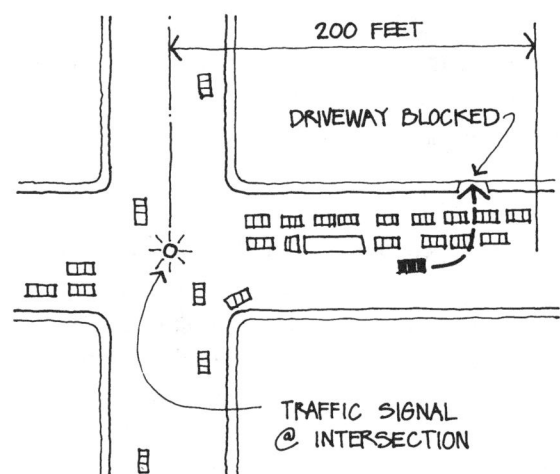

LEFT TURN SITE ACCESS

MINIMUM DISTANCE TO DRIVEWAYS

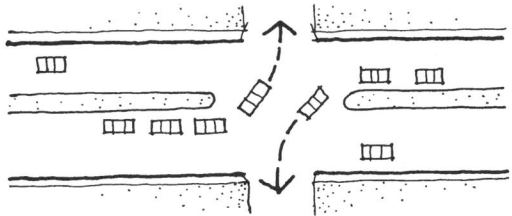

DRIVEWAYS OPPOSITE

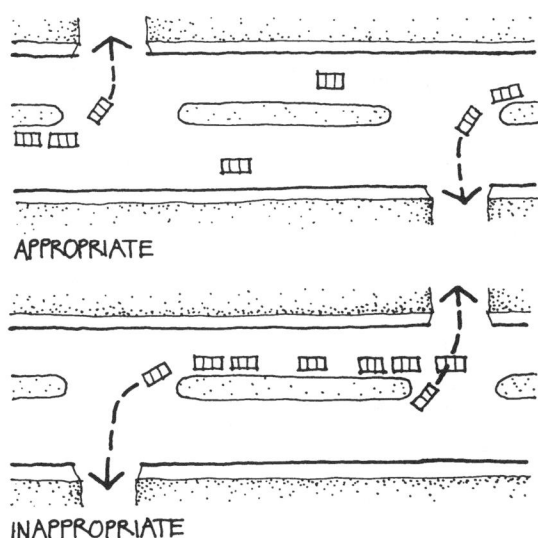

DRIVEWAYS STAGGERED

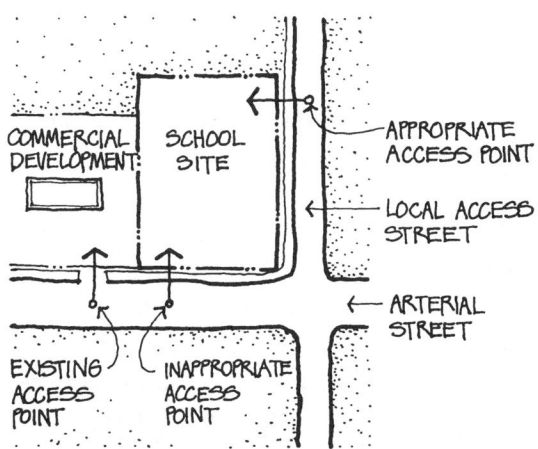

ACCESS - ADJACENT USES

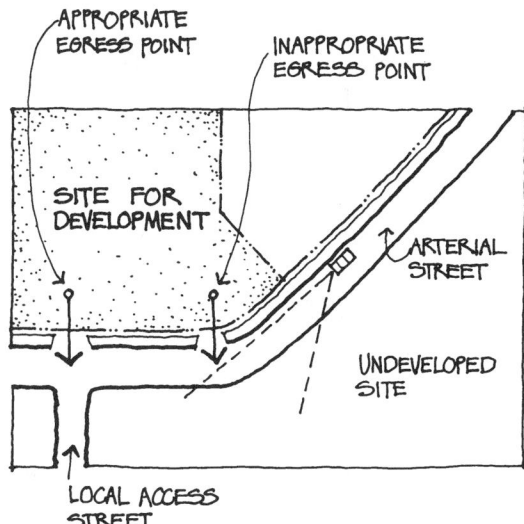

ACCESS - VISIBILITY

The architect must try to solve all of these access problems, or at least provide the best possible compromise solution. On the issue of public safety, however, there are no compromises.

One of the objectives in locating vehicular access points in urban areas is to allow cars to turn either left or right when entering or leaving the site. This is most critical where the amount of street frontage is limited and the site is near a major street intersection.

A sufficient distance between the driveway and the intersection will permit left-turn access onto the site even when cars are stacked and waiting for a signal change at the intersection. The situation improves as this distance increases.

The most undesirable access locations are those close to major intersections, since driveways interfere with traffic moving through these intersections. Cars waiting for a signal change block the driveway entrance when the driveway is less than 200 feet from the intersection. Left-turn access is also blocked during peak periods when waiting traffic backs up past the driveway. Fur-

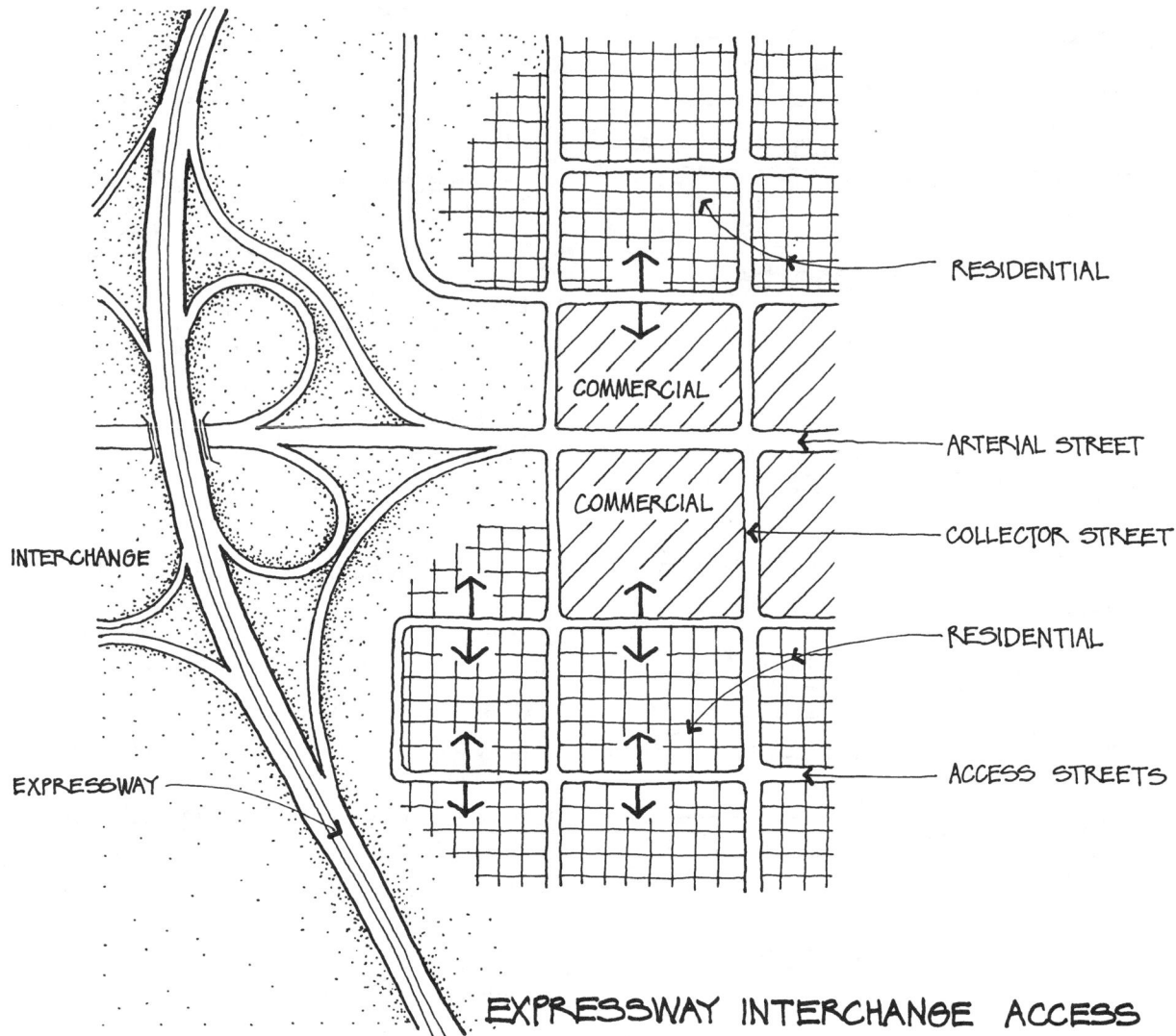

EXPRESSWAY INTERCHANGE ACCESS

thermore, vehicles unable to enter the driveway block the flow of traffic in two directions, causing a virtual standstill. And pedestrians using sidewalks near intersections are exposed to greater potential danger. Access points to sites on opposite sides of streets should avoid interfering with each other. It is preferable to locate driveways directly opposite each other where both share a single access location, similar to a street intersection. Where such an arrangement is not feasible, access points should be offset to provide adequate left-turn stacking space in advance of each driveway and to avoid the overlap of left-turn lanes.

The location of driveways must also consider adjacent developments. For example, if the site is next to an elementary school where children use sidewalks and street crossings, driveways should be located to avoid circulation conflicts, in order to maximize pedestrian safety.

Suburban and rural sites may offer a wider choice for the location of access points. Here, the main consideration might be the preservation of the natural environment. If given a choice, for example, one would not destroy a mature grove of trees to cut a path for roads; one would, more likely, select a route that passes over less valuable land. Furthermore, grading and paving of roads is costly. Consequently, it is always best to locate driveways and align roads in harmony with existing contours. If a great deal of grading is needed because the road opposes the natural topography of the site, the cost will increase, and cutting the natural grade will scar the site's surface and require the construction of berms and banks.

Public safety may be the determining factor in selecting an access point from a highway onto a rural site. The road's alignment and the ability of high-speed traffic to clearly see cars leaving and entering the site may be of greater concern than the convenience of drivers.

When a site is in the vicinity of an expressway interchange, it is preferable to locate access drives to commercial developments along arterial and collector streets at a considerable distance from the interchange. This forces vehicles to reduce speed and allows drivers to become oriented to local traffic patterns and developments. Driveways to residential sites should be from local access streets to further reduce the dangers from traffic.

If, after considering all these factors, there are several suitable access points for vehicles, how does one decide which of these to use? The primary concern should be how these access points relate to the on-site circulation for passenger and service vehicles and to the required amount and optimum location of parking, and how they interface with each other and the buildings on the site.

Parking

Virtually every development requires automobiles to be parked on its site. Most municipalities have ordinances requiring a developer to provide parking based on usable floor area—for example, two parking spaces for every 1,000 square feet of leasable office area. In some cases, the requirements may be stated in terms of parking spaces per unit use, such as two spaces per apartment. Therefore, parking has a major impact on the functional and aesthetic qualities of the site. Whatever the specific requirement may be, parking areas must be designed to minimize ground coverage, provide safe access from public streets, afford ease of circulation for its users, and be functionally related to the buildings they serve.

To move and store vehicles efficiently, one must apply the dimensional and movement characteristics of automobiles to the design of parking areas. Critical factors are the overall length, width, and front and rear overhangs, and the turning radius traced by both front and rear bumpers. Where parking areas are used by trucks and semitrailers, in addition to cars, the design and layout must be sized to accommodate the largest of these vehicles. Turning radii range from about 16'-0" for sub-compact cars to 25'-0" for full-size cars, and 45' to 50' for trucks and semitrailers.

A 9'-0" wide by 20'-0" long parking stall accommodates a full-size car comfortably. However, these dimensions should never be less than 8'-0" wide by 18'-0" long. Parking areas may have a number of different configurations, depending on their capacity, distribution of cars, circulation pattern, and site characteristics. Aisle dimensions are determined by the maneuvering space required to park in each stall and whether they serve one-way or two-way traffic. The width of aisles and parking bays varies, depending on the angle between car and curb. The diagrams shown on the following page indicate parking configurations and aisles, required dimensions, and space needs for various conditions of angled parking.

Parking stalls on either side of an aisle may be at 90 degrees or angled in opposite directions. The latter results in a herringbone pattern and is often used where space limitations prevent perpendicular parking.

In estimating the total area required for parking, it is necessary to include stalls, aisles, and connecting drives. The data shown on the following page are used to compare different parking arrangements and determine a safe and efficient layout. For example, it is more difficult to maneuver a car into a 90-degree parking space than into a 60-degree or 45-degree space. However, the 90-degree layout requires the least amount of area and is less dangerous when backing out because of the wider aisle. Angled parking layouts create one-way traffic aisles, which make parking easier. Angled parking is less efficient, however, because it requires more curb and stall length and creates triangular leftover spaces at each car and at the ends of rows. Where parking aisles are used for drop-off points at building entrances, the layout must consider the additional traffic generated, in order to avoid disrupting the normal traffic flow. Cars should never be required to use public streets or alleys to circulate between parking aisles, nor should they have to back into streets when vacating spaces.

Circulation within parking areas should be continuous. There should be the fewest possible turns, and there should be no dead-end aisles. Parking areas should be sloped to drain, but not more than five percent in either direction, that is, parallel or perpendicular to the direction of parked cars. Slopes should be uniform whenever possible. Driveways leading to parking areas from streets, or ramps connecting separate areas, should also be sloped, to provide a transition between different elevations. Such ramps should not exceed a slope of ten percent.

All layouts must, of course, conform to the requirements for the physically handicapped. Specifications for these are published by the American National Standards Institute, publication ANSI 117.1. Although the number of handicapped stalls is not specified, a reasonable number must be provided in each parking area. These spaces and their adjacent passenger loading zones must be located to provide the shortest possible circulation route to the building entrance.

The minimum dimensions of handicapped parking stalls are 20 feet long and 8 feet wide, with a passenger loading zone no less than 5 feet wide adjacent and parallel to the vehicle pull-up space, for a total width of 13 feet. The five-foot wide loading zone may be located between two handicapped stalls to serve both, resulting in a total width of 21 feet for the two 8-foot stalls and the 5-foot loading zone. An accessible route to the building must be located at the front of the stall, to avoid the hazard of handicapped persons having to circulate behind parked vehicles.

The typical parking layout on the following page shows the handicapped and other criteria for on-site parking of vehicles.

It is preferable to provide separate service drives leading to service points whenever possible. Where service vehicles must share the aisles and drives with passenger vehicles, their routes should be as short as possible to minimize potential conflicts. Loading areas should be out of the way of vehicular routes and provide ample space for service vehicles to reverse and turn.

PARKING DIMENSIONS AND EFFICIENCY RELATIONSHIPS

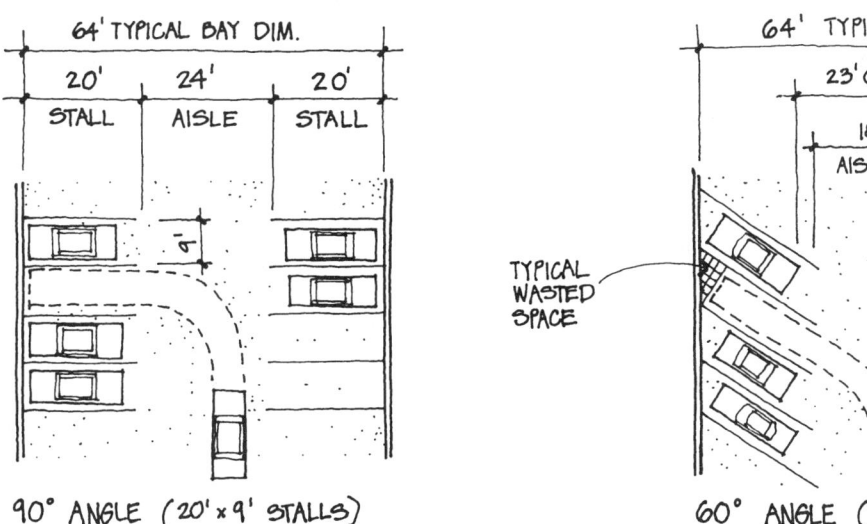

90° ANGLE (20' x 9' STALLS)
- 11.1 CARS FOR EACH 100 LINEAL FT. OF CURB
- 290 SQ. FT. PER CAR AREA REQUIREMENT
- ACCOMMODATES MOST CARS
- PERMITS 2-WAY TRAFFIC AISLES
- MORE DIFFICULT TO MANEUVER

60° ANGLE (20' x 9' STALLS)
- 9.7 CARS FOR EACH 100 LINEAL FT. OF CURB
- 333 SQ. FT. PER CAR AREA REQUIREMENT
- EASY ACCESS
- ONE-WAY TRAFFIC AISLES
- MOST POPULAR CONFIGURATION
- RELATIVELY ECONOMICAL

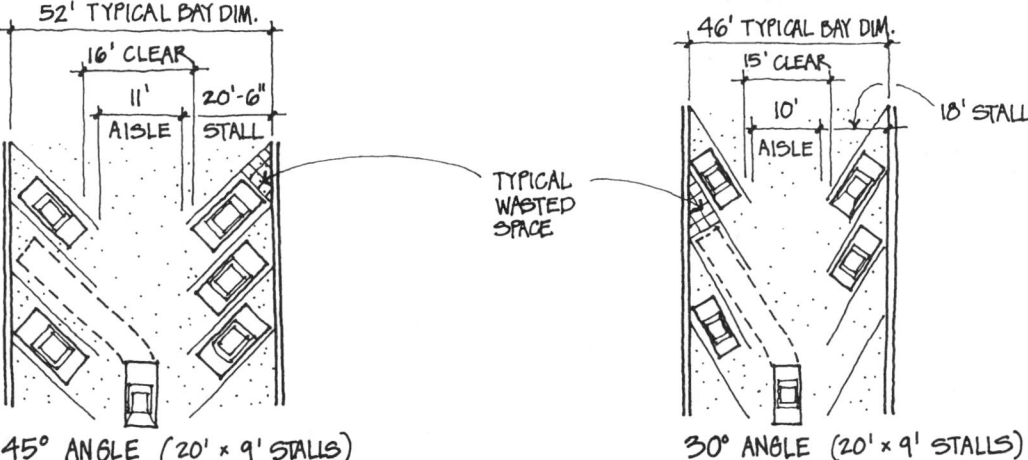

45° ANGLE (20' x 9' STALLS)
- 7.8 CARS FOR EACH 100 LINEAL FT. OF CURB
- 333 SQ. FT. PER CAR AREA REQUIREMENT
- EASY ACCESS
- ONE-WAY TRAFFIC AISLES
- RELATIVELY ECONOMICAL

30° ANGLE (20' x 9' STALLS)
- 5.5 CARS FOR EACH 100 LINEAL FT. OF CURB
- 414 SQ. FT. PER CAR AREA REQUIREMENT
- EASY ACCESS
- ONE-WAY TRAFFIC AISLES
- RELATIVELY UNECONOMICAL

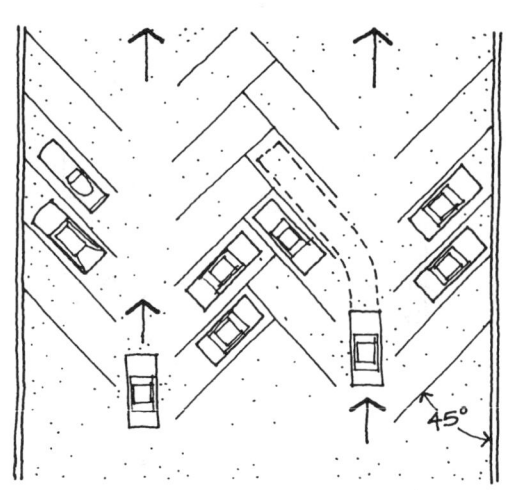
ONE-WAY AISLES; SAME DIRECTION OF TRAVEL IN EACH AISLE.

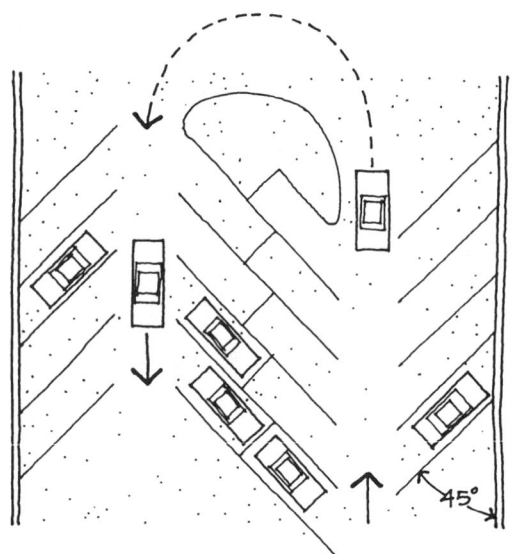
ONE-WAY AISLES; OPPOSITE DIRECTION OF TRAVEL IN ALTERNATE AISLES.

HERRINGBONE PARKING PATTERNS

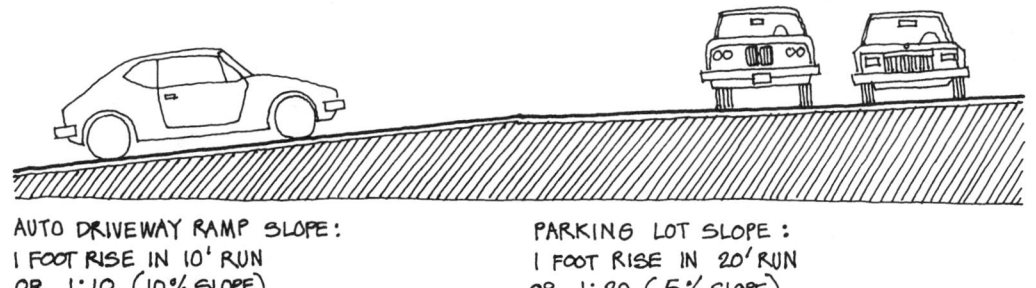

AUTO DRIVEWAY RAMP SLOPE:
1 FOOT RISE IN 10' RUN
OR 1:10 (10% SLOPE)

PARKING LOT SLOPE:
1 FOOT RISE IN 20' RUN
OR 1:20 (5% SLOPE)

MAXIMUM SLOPES FOR VEHICLES

Passengers must be able to circulate safely between automobiles and buildings. Paths may be defined by striping applied to paving, locating raised walkways between bays, and, in heavy traffic situations, pedestrian bridges.

Parking areas used at night should be uniformly illuminated to an intensity of one-half foot candle by regularly spaced lighting standards, 30 to 50 feet in height. Fixtures should be placed to provide an overlap of light patterns and avoid dark areas. The light source for parking lot lighting is usually mercury vapor or high-pressure sodium, for efficiency and economy. The cost of parking includes the cost to acquire the land, grading, paving, curb, drainage, striping, and lighting.

If land costs exceed the cost to build parking structures, it may be more economical to stack layers of cars than to park all of them on grade. An analysis using the average gross area requirement per vehicle, the square foot cost of land, and the unit cost to improve the ground as

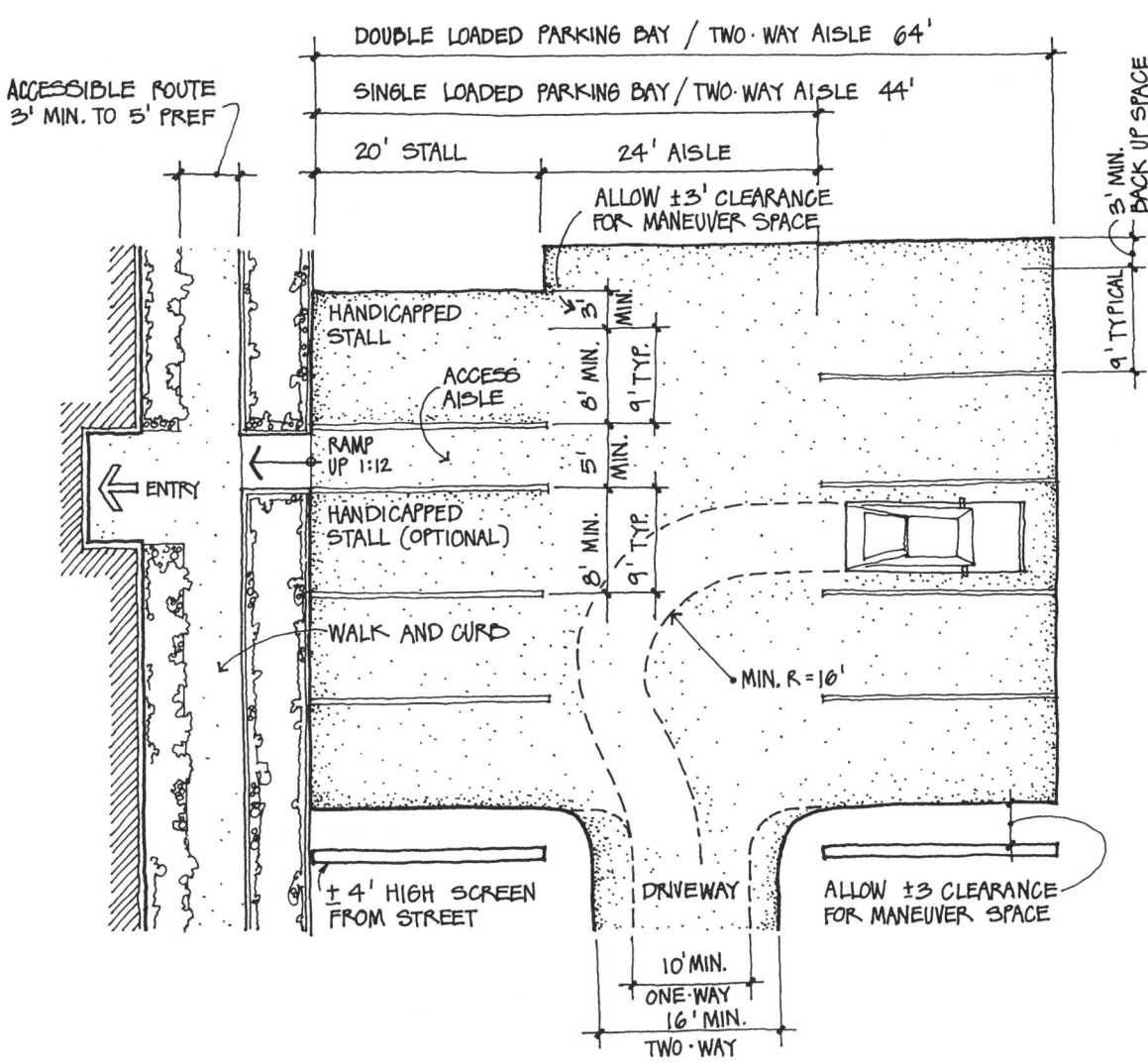

TYPICAL PARKING LAYOUT (PREFERRED DIMENSIONS)

well as to construct parking structures enables one to determine the more cost-effective solution. For example, you might be asked to choose between acquiring sufficient land to accommodate 200 cars on grade or constructing a two-level parking garage on one-half the amount of land, using the cost figures shown in the chart below.

Assuming that the average total area requirement is 350 SF per car, the total area needed for parking is 350 SF × 200 cars = 70,000 SF. To acquire and improve 70,000 SF of land will cost 70,000 SF × $66/SF = $4,620,000. To build a second level of parking would reduce the land requirements by one-half. The cost to accommodate the same number of vehicles for a two-level scheme would be 35,000 SF × $66/SF plus 35,000 SF × $44/SF = $2,310,000 + $1,540,000 = $3,850,000. Consequently, the average cost per car for the parking structure scheme is $3,850,000 ÷ 200 cars = $19,250,

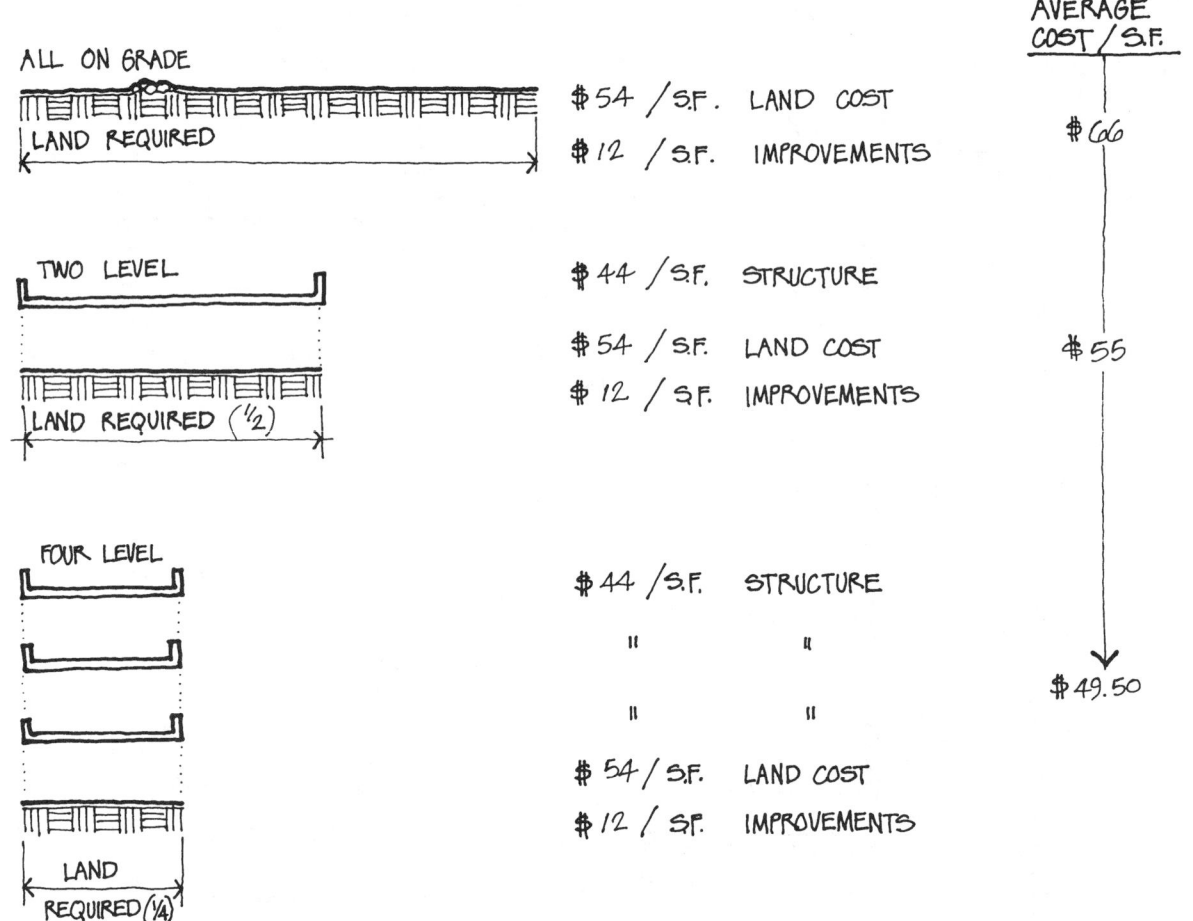

PARKING COSTS

compared to $4,620,000 ÷ 200 cars, or $23,100 per car on grade. As the number of levels of parking structure increases to accommodate more cars, the average cost per car decreases.

Design of Pedestrian Circulation

In downtown locations, building entrances are usually accessible to pedestrians from adjacent public sidewalks. The desire to retain more open space in the cities has prompted developers to sacrifice portions of the buildable area of sites for landscaped plazas, pocket parks, fountains, pools, and similar outdoor areas designed to enhance the environment. Consequently, pedestrian spaces in the city are not necessarily limited to sidewalks.

Furthermore, many cities have barred the automobile from certain commercial zones and created landscaped malls for the pleasure and safety of pedestrians, providing opportunities for social interaction and visual enjoyment. The design of these and other pedestrian facilities requires an understanding of the physical dimensions and movement characteristics of the human body under a variety of circumstances.

For example, the area covered by a human being standing still is approximately three square feet.

This is based on an assumed shoulder breadth of 24 inches and a body depth of 18 inches for the average adult male. In order to move easily in a crowd without creating body contact, a total of 13 square feet per person is required. A lesser area will normally impede movement and require an effort on the part of pedestrians to avoid contact while walking. If the area allowance is less than seven square feet, pedestrians are forced to move as a group rather than to move as individuals unimpeded in all directions. If the area is only three square feet per person, no movement is possible, body contact is inevitable, and a dangerous situation may develop in case of panic caused by a sudden need to reach an exit.

Human motion involves balance, timing, and sight. Uninhibited motion requires spatial allowances for pacing, sensing, and reacting to other pedestrians. Movement on level surfaces differs from that on stairs or ramps, which require more attention to assure pedestrian safety.

The capacity of walkways varies according to the quality and rate of flow. As a rule of thumb, one-way sidewalks should be no less than five feet wide, and collector walks serving large crowds and two-way traffic, from six to ten feet or more, depending on total capacity and rate of flow. Two-directional flow along a walkway is not as efficient as one-way flow. Pedestrians have a tendency to group themselves into nonconflicting directional herds, particularly when the rate of flow is nearly equal in both directions.

Crowd characteristics and psychology also play a role in the design of pedestrian ways. For example, 80,000 football fans streaming out of the Rose Bowl after a game will require greater flexibility of movement in dispersing than 3,000 persons leaving Avery Fisher Hall at the con-

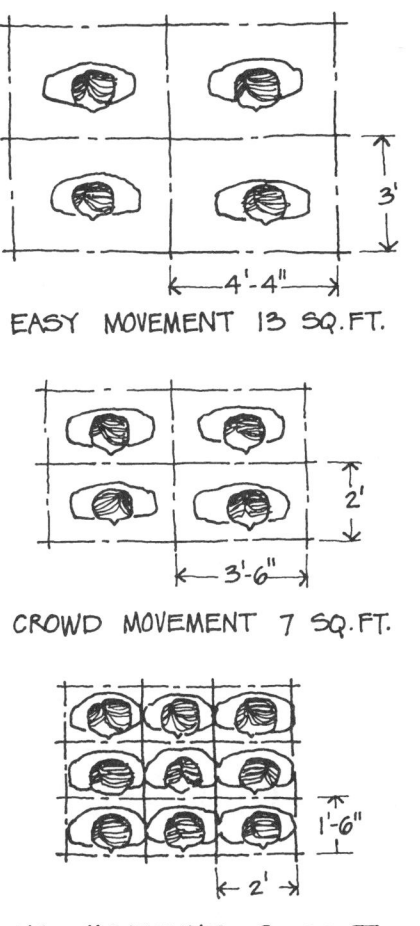

PEDESTRIAN AREA ALLOWANCES

clusion of a concert. Consequently, the designer must allow sufficient space to minimize conflicts which reduce the rate of flow and cause congestion. Each of these aspects must be considered, together with the characteristics and use of the site and its buildings. In a shopping mall, for instance, pedestrians tend to move in linear patterns, criss-crossing diagonally between shops. This requires a greater amount of space to avoid circulation conflicts.

Most people prefer to walk in the line of least resistance. Consequently, foot paths should be

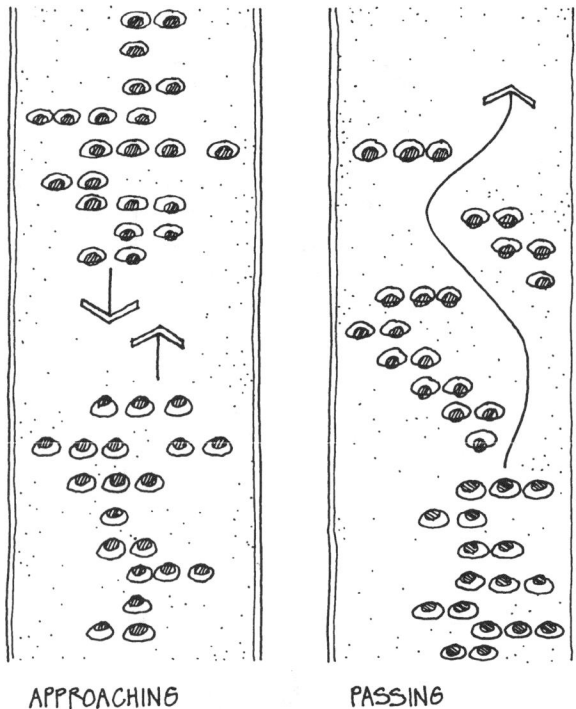

APPROACHING   PASSING

PEDESTRIAN HERD INSTINCT

designed to provide the most efficient and direct means to circulate between uses on a site. In some situations, however, where activities are primarily ceremonial in nature, the configuration of walks leading to major building entrances, monuments, etc. may be determined more by formal than by functional needs, as at the Capitol Mall in Washington, D.C.

Distance and speed are the major limitations in the design of a site's pedestrian circulation system. Most people are unwilling to walk more than half a mile in performing their routine tasks. Since the average walking speed is about two and one-half miles per hour, pedestrians will walk up to twelve minutes. If the distance is more than half a mile, people prefer to use their cars to reach their destinations.

Many of our older universities, for example, have grown in size beyond the expectations of their founders. Where distances are excessive, students often prefer to drive between facilities, though they may have to walk five or ten minutes to and from their car, leave and re-enter the site, and thereby use more time than the entire trip might require on foot. To avoid this, site planners concentrate high-use facilities within a core comprising a walking radius of one-quarter mile, which circumscribes an area of approximately 125 acres. Low-use facilities, especially those for parking and service, are placed around the perimeter of the core. This reduces walking time within the core, allowing one to traverse the area in about 12 minutes from end to end, and minimizes vehicle-pedestrian conflicts as well.

Walks should be designed to allow people to move safely, independently, and unhindered through the exterior environment. The surfaces of walks should be stable and firm, relatively smooth in texture, and have a nonslip quality.

Walking surfaces can be grouped into three types: soft, variable, and hard. Each group has its advantages and disadvantages. The soft surfaces are usually the least expensive initially, but require a high degree of maintenance. They are susceptible to erosion, can withstand only light traffic, and are difficult for handicapped people to walk on. Soft surfaces, such as crushed rock, earth, lawn, river rock, soil cement, and tanbark, are useful for areas where light pedestrian traffic needs a moderately firm surface, such as recreation areas, parks, and nature areas.

Variable surfaces, such as cobblestones, exposed aggregate, flagstone, sand-laid brick, wood decking, and wood disks in sand have moderate maintenance requirements and moderate to high installation costs. The irregularity

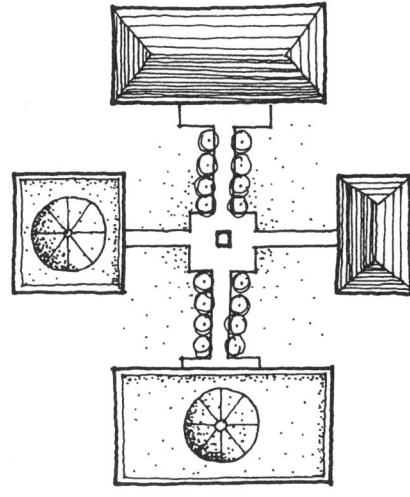

FORMAL

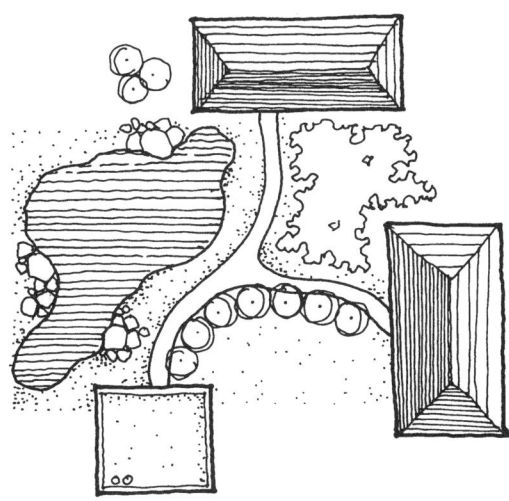

FUNCTIONAL

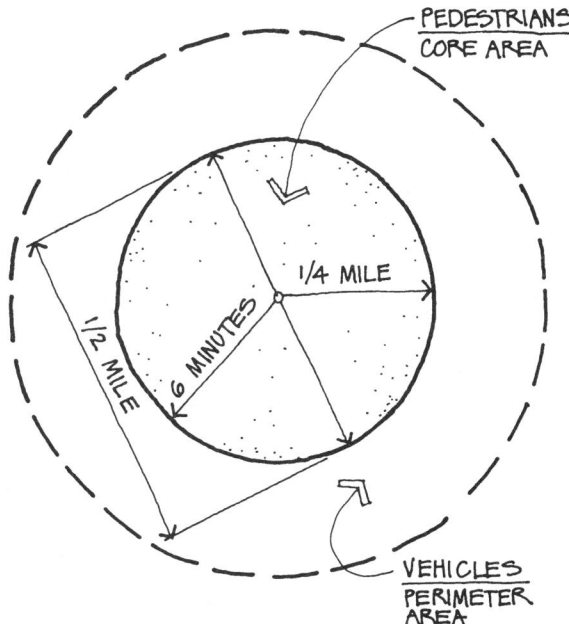

PEDESTRIAN DISTANCE AND SPEED

Hard surfaces, such as asphalt, concrete, and tile or brick in concrete, are usually the most expensive initially, but their maintenance costs are relatively low. They provide smooth, firm, and regular surfaces for walking or moving wheeled vehicles, including wheelchairs. Expansion and contraction joints, which are less than 1/2" wide and filled, are kept to a minimum, and snow and ice can be removed without extensive damage to surfaces.

The Sustainable Design philosophy promotes permeable parking surfaces which allow percolation of water directly on site rather than forcing liquids off the parking area. Permeable parking can be designed with grassy infill which has less heat build-up and is visually more attractive. Winter snow removal of permeable surfaces should be done carefully with a hard rubber (not steel) plow blade.

Pedestrian paths with gradients under 5 percent (1:20) are considered walks. Those with gradients in excess of 5 percent are considered ramps and have special design requirements. Routes of their surfaces and their wide joints make walking difficult for the handicapped. Furthermore, ice and snow can damage the surfaces or be difficult to remove.

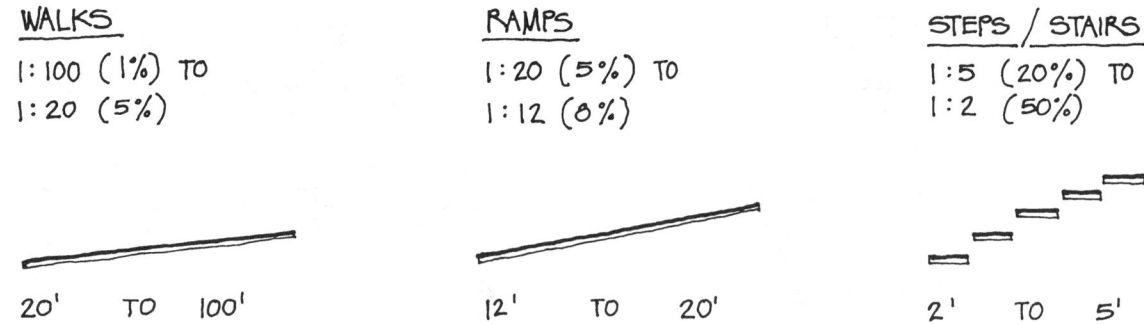

REQUIRED DISTANCES TO RISE OR DROP ONE FOOT VERTICALLY

with gradients up to 5 percent can be negotiated independently by the average wheelchair user, but sustained grades of 4 to 5 percent should have five-foot level areas every 100 feet of run or so to allow a handicapped person to stop and rest. However, slopes under 3 percent are preferred, whenever possible.

Walkway lighting may vary from one-half to five footcandles, depending on the intensity of pedestrian use, the hazards present, and the relative need for personal safety.

Changes in grade from street to sidewalk and from sidewalk to building entrances create problems for handicapped persons. Curb ramps not exceeding a gradient of 1:12 (8 percent) facilitate movement over low barriers. Their surfaces should be nonslip but not corrugated, since frozen water in the grooves can cause ramps to become slippery. Although curbs may be required in certain situations, they should not be higher than 6" to 6-1/2". Curb ramps for the handicapped should be provided wherever pedestrian traffic occurs and where vehicles are parked adjacent to curbs.

Walkways should be pitched laterally about 1/4 inch per foot to eliminate standing water in depressions. For wider walks, the surface should be crowned to slope away from the center.

Stairs may be used where gradients are such that persons in wheelchairs must use an elevator to negotiate the change in level. An exterior stair should have at least three risers in order to avoid pedestrians tripping over steps not easily seen. Stairs with more than four risers should be provided with handrails on at least one side. Exterior stairs typically have 5-1/2" to 6" risers and 15" and 14" treads, respectively. Broad expanses of monumental steps may have 3" to 4" risers with 19" to 17" treads respectively. These should be avoided where pedestrian circulation is heavy, since they are more cumbersome than normal stairs. A useful rule for proportioning conventional steps is that the height of two risers added to the tread dimension should equal 26" to 27". Stairs in heavy public use should never have a gradient over 50 percent, for example, a 6" riser with a 12" tread.

Pedestrian spaces must be sized according to the anticipated intensity of activity. For example, a campus mall in a central location may appear oppressive because of the crowd attending an outdoor concert, while it would seem empty, vast, and lonely in the evening, when most

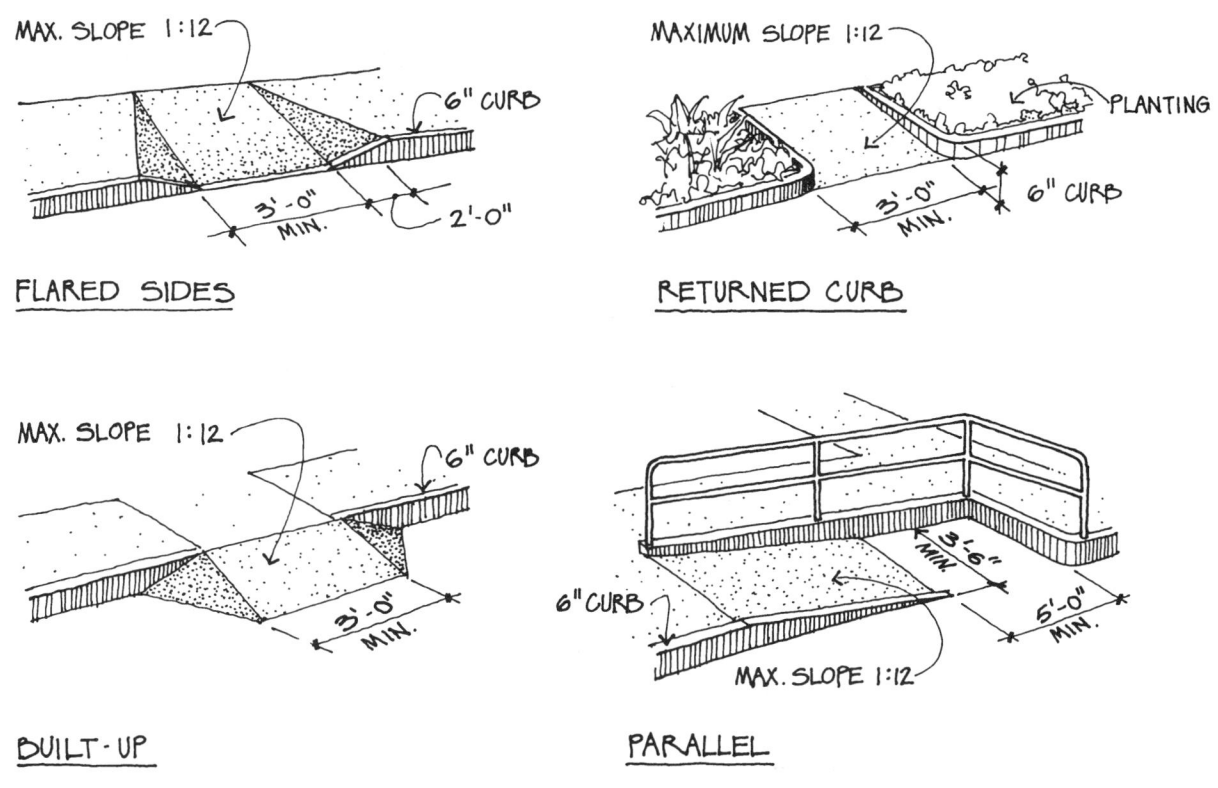

TYPICAL CURB RAMPS

students are no longer on campus. A foot path on the periphery of a commercial center may seem excessively long, while if it were bordered by varied shops and stores and used by many people, it would appear interesting and relatively short. Pedestrian spaces should do more than physically link activities—they must reinforce the composition of the site design, both visually and emotionally.

ZONING

General

The first American zoning ordinance was enacted in New York City in 1916 to limit the size and shape of new skyscrapers so that the adjacent streets would not become permanently shaded canyons. Most zoning statutes in the 1920s dealt with physical development. They divided cities into districts of different uses, with uniform regulations for each. For example, residential districts permitted only residences, commercial districts only commercial activities, and so on. Major categories were further divided; for example, industry was divided into heavy manufacturing and light manufacturing. Residential use was divided into single-family, two-family, and multiple-family dwellings. Homogeneous districts were based on the idea that differing uses within a district would lower property values. Multi-use districts, sometimes called *cumulative zoning*, allowed residences in commercial zones and residential and commercial uses in industrial zones.

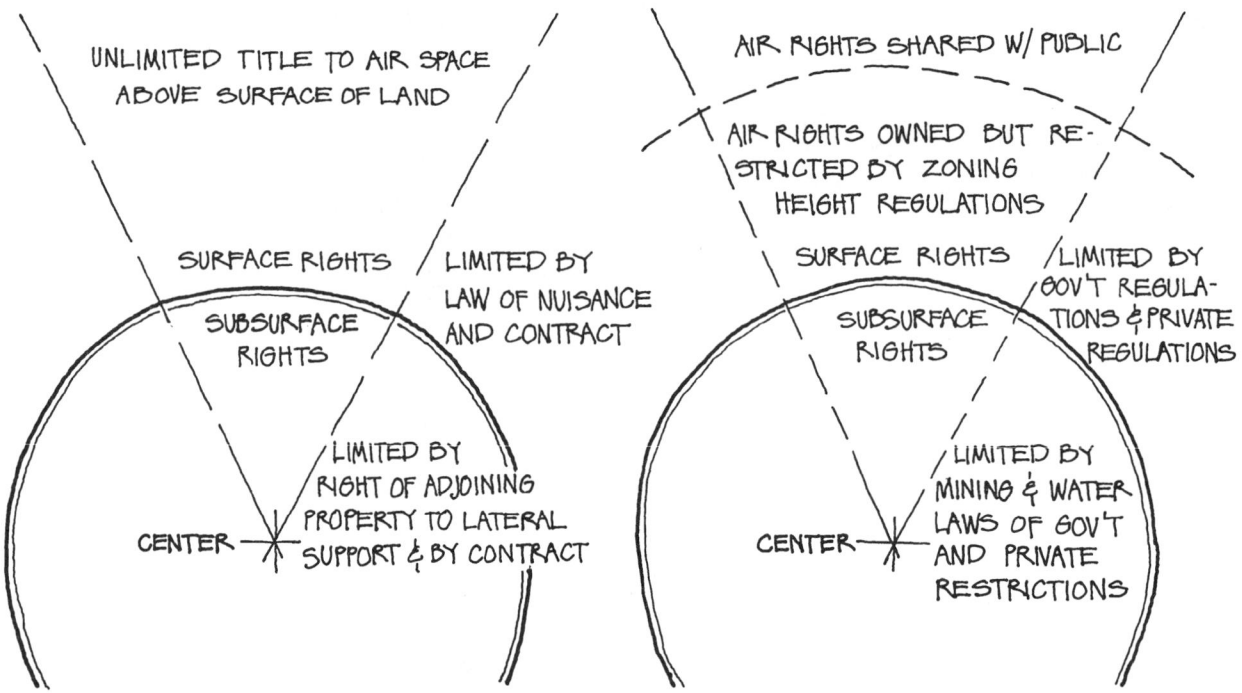

PROPERTY RIGHTS AFTER REVOLUTION AND BREAK FROM ENGLAND

PRESENT PROPERTY RIGHTS

The ordinances of the 1920s also regulated the height and bulk of buildings and setback lines. The intent of these acts was to allow the owner to develop the land as he or she wished, as long as the specific restrictions of the ordinances were not violated. These laws authorized, but did not compel, local authorities to control development decisions. They did not offer incentives to owners to undertake desirable development, but rather, attempted to avoid undesirable development.

Since that time, the ordinances have changed considerably; instead of prohibiting poor planning, they now encourage, and sometimes even compel, desirable planning.

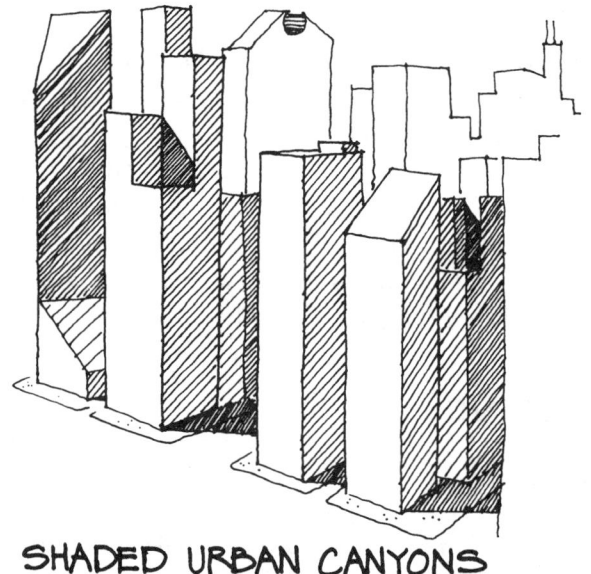

SHADED URBAN CANYONS

Lesson Five: Site Planning Issues: Designing the Site   **77**

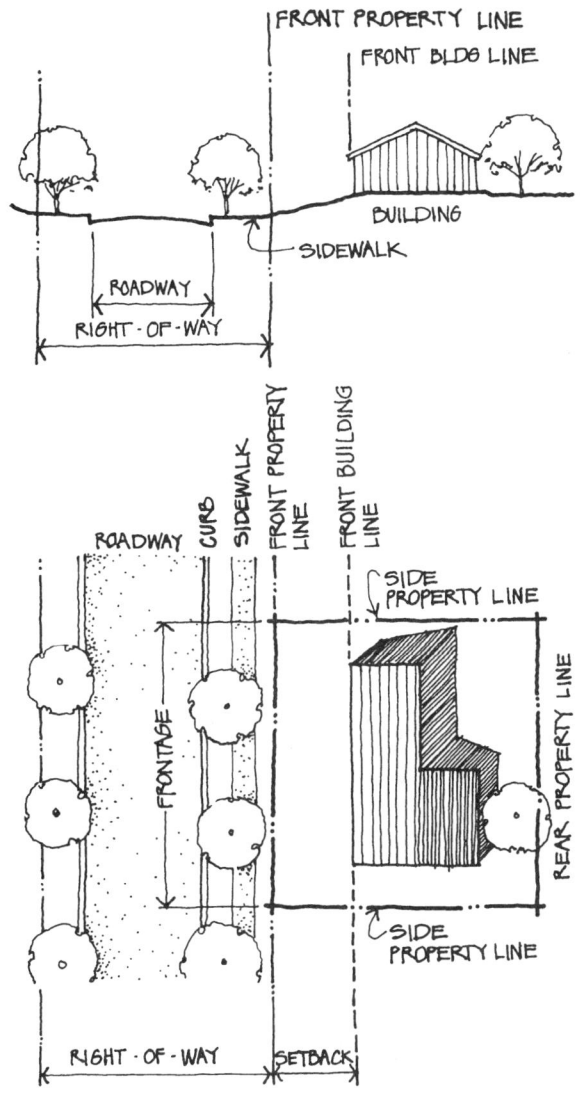

TYPICAL DEVELOPMENT STANDARDS

The enactment of the Model Land Development Code recognized aesthetics, environmental problems, and the preservation of historical sites as planning and development factors. In New York City, developers commonly add plazas at the ground level of office towers, in return for permission to erect taller buildings. For example, IBM was allowed to build five additional floors in return for creating a tree-filled atrium at the foot of its Madison Avenue building.

Zoning ordinances often restrict the height and size of buildings, as well as their location on the site. They may prescribe setbacks from property lines, limit percentage coverage of the lot area, restrict the number of dwellings per acre (density), and require a specific amount of off-street parking, as well as numerous other possible regulations.

Zoning Envelope

The volume within which a building may be placed is sometimes referred to as the *zoning envelope*. This is an imaginary, tent-like space inside of which the building may be placed in any location, so long as it does not penetrate any of the imaginary surfaces. The drawing below depicts a highly simplified zoning envelope. Most of these have more complex shapes determined by several interacting standards and restrictions.

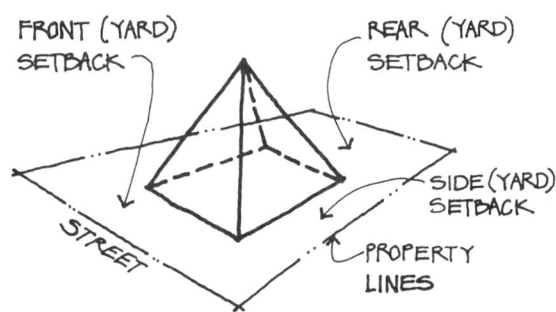

THE ZONING ENVELOPE

Setbacks and Yards

A zoning ordinance often regulates the distance between a street and a building, as well as between buildings. The main purposes of such restrictions are to provide building interiors with natural light and ventilation, inhibit the spread

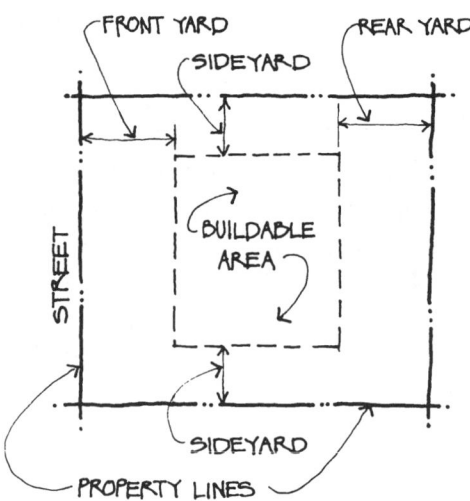

TYPICAL YARDS
SINGLE FAMILY DISTRICT

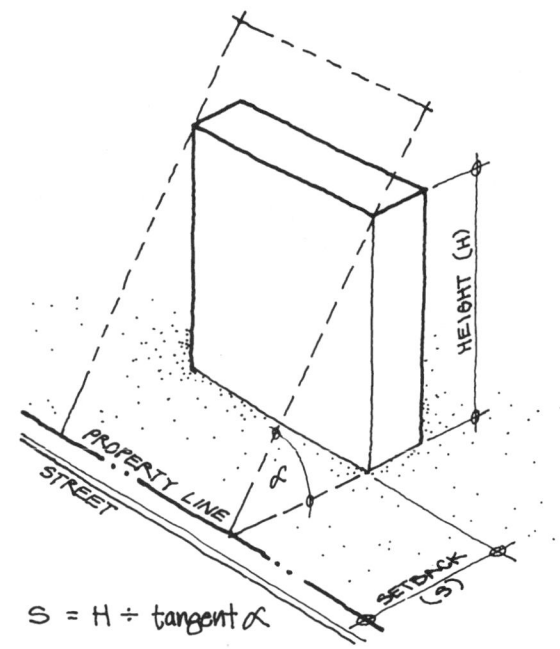

$S = H \div \tan \alpha$

SETBACK AS
A FUNCTION OF HEIGHT

of fire from one structure to the next, and minimize conflicts between street traffic and off-street activities. These ordinances also allow for future street widening and create space open from the ground to the sky. The regulations for yards and other setbacks establish the base of the zoning envelope (the ground area within which construction may occur). The drawing above shows a lot for a single-family house and the various yards referred to in zoning ordinances.

A setback is the horizontal space adjacent to a property line into which a structure may not project. Setbacks provide a sense of openness, as well as light and air. They may also be required for off-street parking, or they may be primarily for aesthetic reasons. In some instances, setbacks are established as a function of the building's height using a formula which requires taller buildings to be further back from property lines, in order to assure a minimum amount of openness to the sky. Most ordinances allow paved parking areas within setbacks; however, they may require trees at specific intervals or low walls at property lines in order to screen the rows of parked automobiles.

Height Limitations and Variable Setbacks

Zoning ordinances may limit the number of stories in a building, its height in feet above street level, or both. The height is usually measured from grade, which may be defined in various ways, depending on the local zoning ordinance. For example, grade may be defined as the lowest adjacent ground elevation, the average adjacent ground elevation, or perhaps some other elevation. Thus, the way in which grade is defined may affect the number of stories permitted, particularly on a sloping site.

Some zoning ordinances contain variable height and setback requirements, either in place of

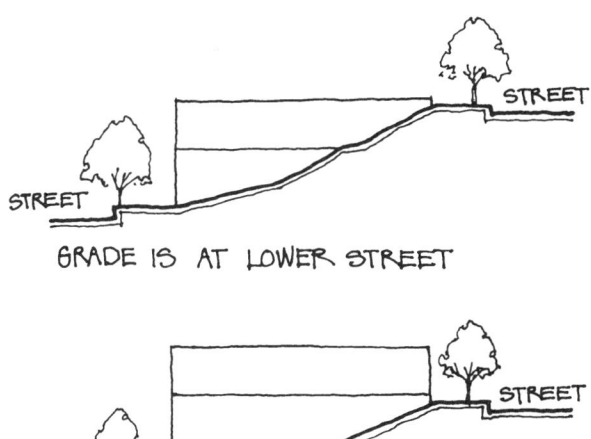

ALTERNATE INTERPETATION OF GRADE

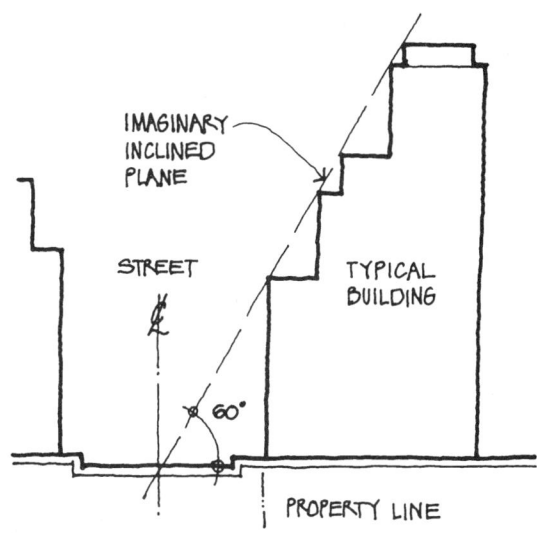

SETBACK BASED ON INCLINED PLANE

absolute limits, or in addition to them. For example, in the diagram shown, no portion of a building may be placed closer to the street than an imaginary plane inclined at an angle of 60 degrees with the street and extending upward from the center line of the street. A control of this sort encourages a "ziggurat" building profile.

Height limitations are more common in residential zones than in commercial or industrial zones. Crowded urban areas, such as New York City, limit the number of stories in order to control the population density and the resulting traffic, and to retain a certain amount of sky exposure.

Land Coverage

In addition to the basic zoning envelope, there are other restrictions which determine how large a building may be placed on a site. These controls, which involve the amount of development, limit the proportion of the site that may be covered by buildings (land coverage) and the ratio of total usable floor space to total site area (floor area ratio or FAR). Land coverage is expressed as a maximum percentage of total available land area that may be covered by a building or buildings. The open land may be used for surface developments, such as parking, plazas, recreation and other types of landscaped spaces, and level or depressed courtyards. The purpose of these restrictions is to encourage the retention and development of open spaces, to enhance the environment through the admittance of light and air, and to provide planted areas to relieve the hard surfaces of buildings, sidewalks, and streets.

Floor Area Ratios

The floor area ratio (FAR) is the ratio of the floor area of a building to the total area of the site. The purpose of an FAR ordinance is to control the amount of site development and to restrict the bulk of a building, in order to encourage openness, light, and air, especially in urban areas.

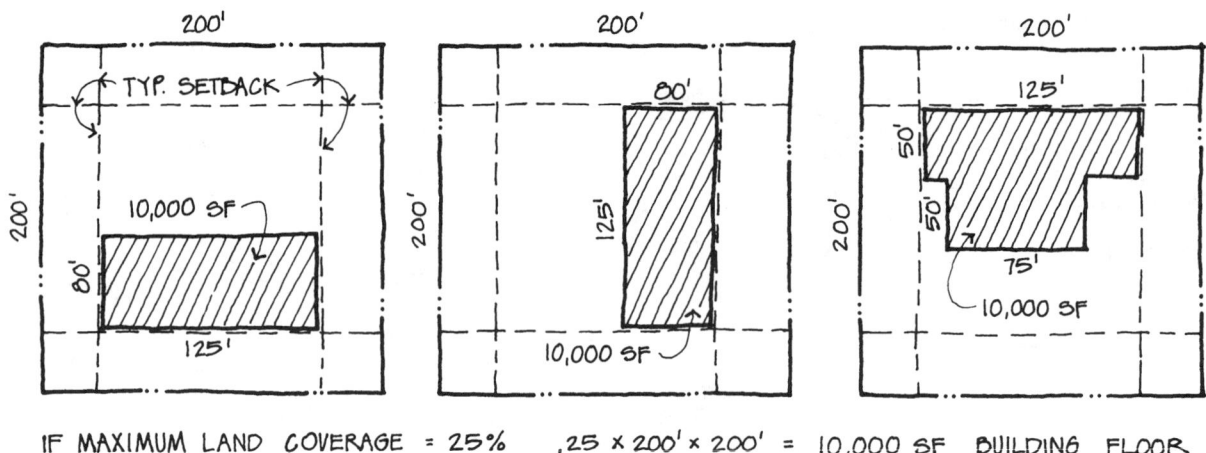

THREE OPTIONS FOR LAND COVERAGE

Thus, a floor area ratio of 2.0 would permit 40,000 square feet of floor space on a site of 20,000 square feet (2 × 20,000). A ratio of 2.5 would permit 50,000 square feet of floor space on the same site. The drawings on the following page show three options that would be available to the owner of a lot having a maximum FAR of 4. In each of the three illustrations, the lot size is the same: 100,000 square feet. Thus in each case, the maximum floor area allowed is 400,000 square feet. In figure A, the owner has covered the entire site with a four-story structure containing 100,000 square feet per floor. In figure B, half of the site has been covered. Since each floor has 50,000 square feet, the building can be eight stories. In figure C, the entire site is covered with a one-story structure, using up 100,000 of the 400,000 square feet of floor area allowed, and the remaining 300,000 square feet have been put into a twelve-story tower, with each level containing 25,000 square feet of usable floor area.

In all of these examples, floor area may be either the net usable space, excluding stairs, elevators, and other similar spaces, or the total gross area of the building, depending on the ordinance.

Off-Street Requirements

Many cities require an owner to provide a minimum number of off-street parking spaces for a building's tenants and visitors. For residential buildings these requirements are usually expressed in terms of parking spaces per dwelling unit. If dwellings are small, one space may be required; for larger units, however, two spaces are normally a minimum. For office buildings in commercial zones, the requirement is usually stated as one space for a specified amount of usable floor area. For example, if the requirement is one parking space for every 500 square feet of usable floor area and the building contains 50,000 square feet, the owner must provide parking facilities for no less than 100 cars (50,000 ÷ 500) within the limits of his or her property. In some cases where space is limited, it may be possible to satisfy the parking requirement on a separate site, provided it is located within a prescribed distance from the building site. Some districts also require a certain amount

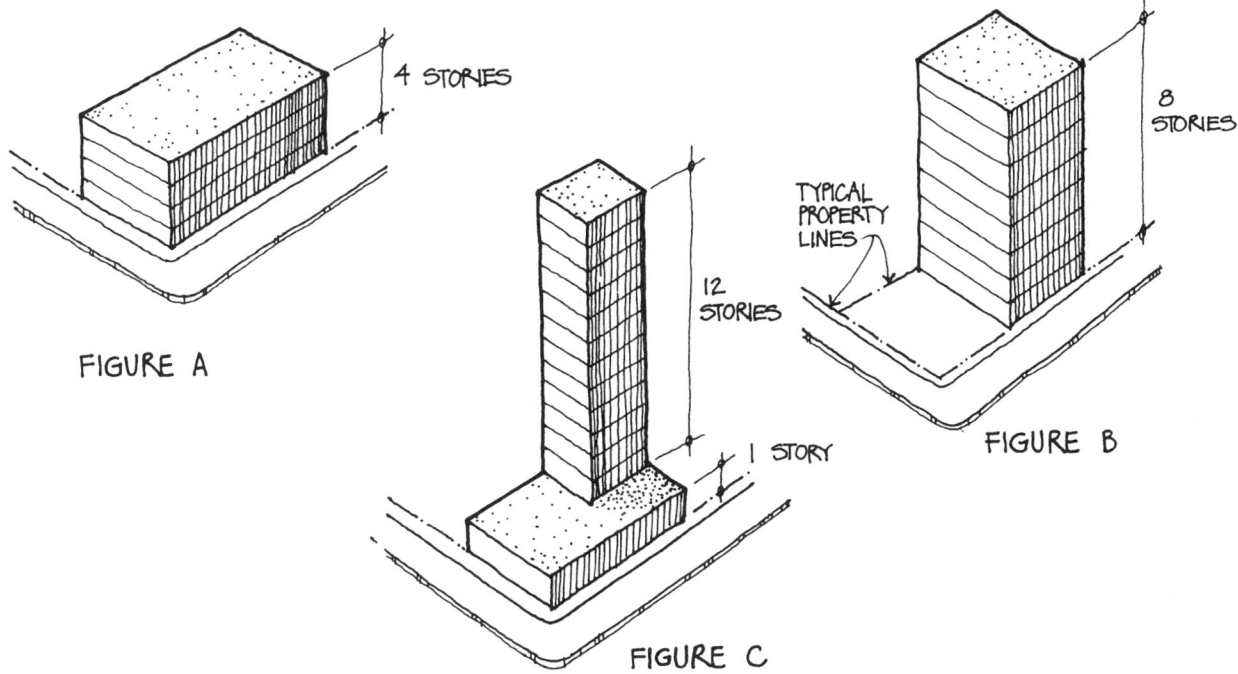

THREE OPTIONS WHERE FLOOR·AREA·RATIO IS 4.0/LOT IS 100,000 S.F.

of site area for loading and unloading of service vehicles, particularly in the case of hospitals, hotels, and institutions.

FLEXIBLE ZONING

General

The purpose of flexible zoning is to overcome the rigidity of traditional zoning and to make the regulations relevant to changing patterns of development. Most zoning ordinances continue to reflect the basic principles of the traditional ordinances of the 1920s: regulations which rigidly define the way land may be used and limitations concerning its physical development. More recently, certain modifications have been introduced which make zoning ordinances more flexible, while preserving their intent.

The *conditional use*, for example, is a departure from traditional zoning, which prohibited any uses in a district other than those specifically allowed by the ordinance. Other significant deviations from traditional zoning concepts include the *planned unit development* (cluster concept), the *floating zone, incentive (bonus) zoning*, and *contract zoning*. Together, these devices are sometimes called *flexible zoning*.

Variances and Conditional Uses

Because even the best zoning ordinances may cause an unintentional hardship to owners of specific land parcels, most cities have established boards which have the authority to grant exceptions to or deviations from the precise terms of these ordinances. These exceptions are called *variances*. Theoretically, a variance is granted when the literal application of an ordinance would cause an undue hardship in the proposed development of a site. For example, a site may have a width of 280 feet in an area where the zoning ordinance specifies the minimum to be 300 feet; however, the site exceeds the minimum lot size by 20 percent. Under these circumstances, a zoning board might grant a variance reducing the lot width requirement because the property conforms generally to the intent of the law.

Conversely, if all existing buildings along one side of a street are set back 20 feet from the front property line, a zoning board would be reluctant to grant a variance allowing an owner to build up to the property line, because this would create a detrimental visual contrast, easily perceived by the surrounding property owners and the public.

The purpose of a *conditional use* in a zoning ordinance is to provide for flexibility within a district. If a use is described in an ordinance as a "conditional use," it is permitted only if specified conditions are met, a public hearing has been held, and approval has been given by the local governing body. A conditional use is normally granted if it is considered to be in the public interest. A school serving local residents, for example, may be permitted in a residential zone, providing it conforms to certain criteria for traffic, pedestrian walks and crosswalks, and noise control. Under such circumstances a zoning board may grant the conditional use of a site subject to restrictions for the protection of adjacent property owners. The granting of a conditional use or special-use permit does not, however, change the zoning of the particular parcel of land. If the development is abandoned, the conditional use would no longer apply and the property would revert to its original district designation.

Rezoning

The only alternative available to a landowner who cannot meet the requirements for a conditional use permit is to seek rezoning of his or her property. Rezoning, however, can cause hardships to neighboring property owners. Under the strict interpretation of a zoning ordinance, a choice may have to be made between the interests of the landowner and those of the neighboring property owners. Rezoning small individual lots results in *spot zoning*, which may alleviate an owner's hardships. There are times, however, when rezoning is accomplished through political manipulation, rather than for legitimate reasons. For example, rezoning a parcel of property from residential to commercial might appear to have a reasonable basis, while the real reason may be to increase the value of the property for the benefit of its owner.

Contract Zoning

An agreement between a developer and local government to restrict usage or height, or to provide additional setbacks or buffers, over and above what is required by the ordinance, in return for certain benefits, is called contract zoning. For example, a developer may agree to additional restrictions in return for being granted approval of a conditional use. Such restrictions exceed the requirements of the local ordinance and are legally binding.

Contract zoning gives the local governing body power to rezone land or issue special permits granting permission to develop land for nonconforming uses in exchange for a developer's commitment to perform certain compensating acts. In San Francisco, for example, a developer must pay for the construction or renovation of a certain number of housing units whenever he or she plans to erect a new office building. The purpose is to create new dwellings for office

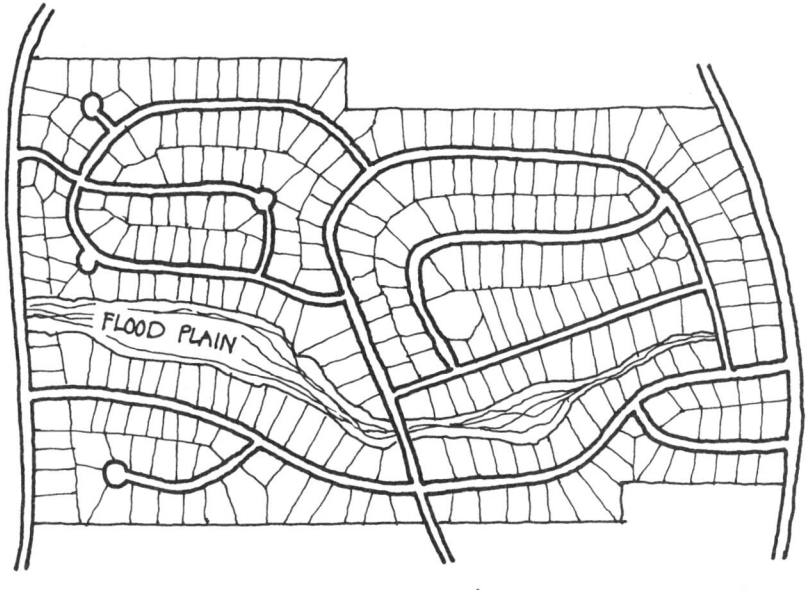

CONVENTIONAL DEVELOPMENT / TRADITIONAL ZONING

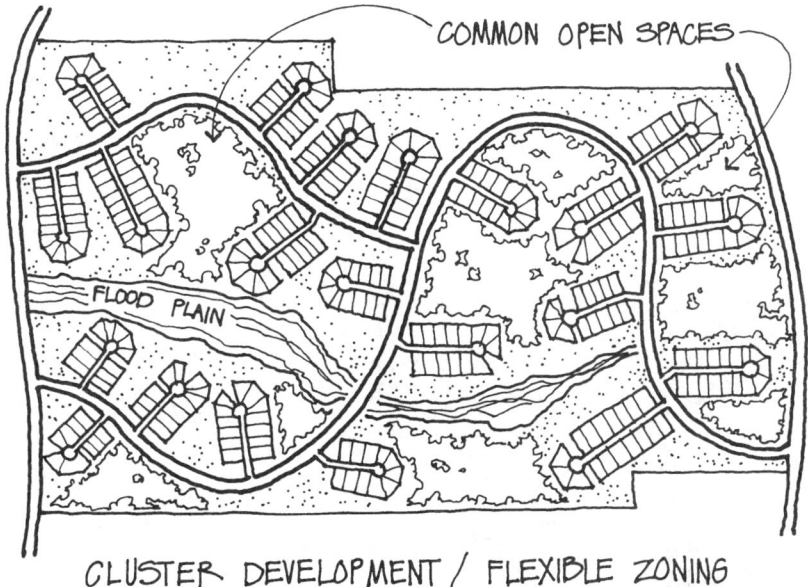

CLUSTER DEVELOPMENT / FLEXIBLE ZONING

FLEXIBLE ZONING

workers in an already crowded area with housing shortages. Should the developer balk, he or she must contribute a certain amount of money to a housing bond program, based on the area of new office construction. Other promised acts or conditions might include noise abatement, traffic control, or the erection of walls and landscaped buffer zones. As an alternative, the developer may be allowed to contribute the necessary funds to the local authority in lieu of performing the work himself.

Bonus or Incentive Zoning

Traditional zoning is inflexible in the sense that owners are forbidden to develop their property contrary to the zoning ordinance. Traditional zoning, therefore, prevents the worst from happening, but that is all it can do. It cannot assure good planning since it is only a restraint. Architects have opposed the negative aspects of zoning while searching for ways to make it a more positive force in development. Gradually, ordinances have been modified in order to reward builders for benefiting communities.

In some cities, zoning requirements may be waived if the developer provides bonus features, as in the IBM Building in New York. This is often attractive to the developer, because not only can the floor area or height of a building be increased, for example, but the bonus features may provide amenities which make the project more desirable for tenants, thereby increasing rents.

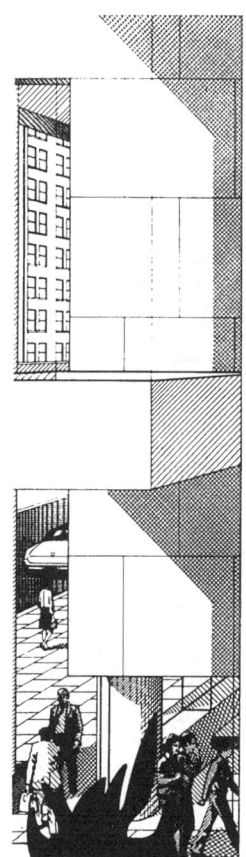

INCENTIVE ZONING:
A STREET-LEVEL ART GALLERY
IN EXCHANGE FOR A LARGER F.A.R.

Incentives can be given for a variety of reasons: street widenings, providing unobstructed views (as along a shore line), inclusion of theaters and retail space in office buildings, provision of walkways for public use (such as a pedestrian bridge over a street), and preservation of open space. Since open space is the most common objective of incentive zoning, allowing a greater floor area ratio is probably the most prevalent incentive. For example, the developers of the Bankers Trust Building in Manhattan were allowed greater tower height and floor area in exchange for providing a large elevated open plaza and a two-level covered arcade of shops.

LESSON SIX
SITE DESIGN VIGNETTE

Introduction
Vignette Information
Design Procedure
Analyzing the Site Plan
Important Parking Objectives
Arranging the Elements
Circulation
Final Arrangement
Key Computer Tools
Vignette 1 Site Design
    Introduction
    The Exam Sheet
    Site Analysis
    Bubble Diagram
    Circulation
    Diagrammatic Arrangements
    Potential Problems
    Final Site Design

INTRODUCTION

The Site Design vignette is the longest of the three problems that comprise the Site Planning division. As discussed in the introduction, NCARB recently combined the one-hour Site Design vignette with the thirty-minute Site Parking vignette. Site Design is a comprehensive exercise that tests your understanding of general site planning principles, including parking, building placement within building limits of the site, and site circulation. Candidates must develop a site plan based on a program that includes a number of specific building elements, site influences, parking and circulation requirements, and legal restrictions. You must integrate these programmatic demands, code restrictions, vehicular site access issues, parking requirements, and environmental considerations into a workable and logical design. Solutions will be analyzed for compliance with program requirements, completeness, and technical accuracy.

VIGNETTE INFORMATION

The Site Design vignette begins with an index screen, which offers the choice of several additional screens containing necessary information. Among these are:

- **General test directions**—This general information appears on the index screen of every Site Planning problem. Since it applies to all vignettes, candidates need not refer to it more than once. The screen includes the following advice:

    1. Read all directions and become familiar with the scope and nature of the problem.
    2. Ignore all other codes or standards that may be in conflict.
    3. Make no assumption of conditions unless they are specifically stated.
    4. Your initial solution may be developed on screen with computer tools or on scratch paper.

5. No reference material is permitted other than what is on the screen.
6. No paper may be used other than that provided.

- **Vignette Directions**—These offer instructions about preparing your site design. It describes the site on which to place the buildings, parking lot, related site elements, and vegetation. One is advised that the location of all elements must relate to the existing environmental conditions. The programmed elements are listed as follows:

  1. Buildings
  2. Outdoor space
  3. Parking spaces
  4. Pedestrian walkways
  5. Vehicular access and service drive
  6. Vegetation

- **Program**—This screen contains a description of each required element, including specific adjacencies, orientation, and restrictions. The height of the buildings and size of the open space are specified, as is the number of required car spaces. Other data include orientation regarding sun, wind, and views, vehicular and pedestrian circulation requirements, and any other detail affecting the layout, such as the number of permitted curb cuts.

- **Tree Diagrams**—Illustrated here are the diameter and height (in plan and elevation) of both deciduous and coniferous trees that may appear on the given site.

- **Tips**—These are suggestions, generally regarding specific computer tools that help the candidate work more effectively. For instance, use the *move group* tool to move a bank of parking spaces, and make minor adjustments to elements with the *zoom* tool. The tips area might also suggest that parking spaces be laid out to the proper scale, the dimensions of which may be read directly in the *element information* area of the screen. You can then rotate or reposition spaces with the *rotate* tool or *move group* tool. Other useful tools for the Site Design vignette that this screen might highlight include the *move, adjust* tool to modify the position or size of drives or sidewalks, and the *sketch line* tool to layout clearances. A final helpful tip that the this screen will most likely emphasize is that road and drives are joined by connecting their dashed centerlines.

The work screen, on which your solution is actually presented, is displayed by pressing the space bar on the computer keyboard. One may toggle between this screen and any other screen in the same way. The work screen contains the site plan that shows existing roads, property lines, building limit lines, easements, trees and other natural features, wind direction, and north arrow. Site contours are not indicated, since topography is not a part of the Site Design problem. Along the left side of this screen are found the computer tool icons, including the draw tool, which brings up the list of required elements.

DESIGN PROCEDURE

The Site Design vignette requires candidates to arrange in plan a number of elements that conform to the programmatic requirements. The resulting solution will be a schematic plan, not unlike those produced each day in most architectural offices. Diagrams such as these are simple in appearance, and most can be developed in a relatively short time. Buildings are represented by simple rectangles, and paved open spaces, roads, and parking areas are shown as simple geometric shapes defined by single solid lines.

Some specific hints regarding the parking area portion of this vignette will be helpful. Before laying out the actual parking spaces, have an idea of where on the site the parking lot will be located. In this regard, a quick pencil sketch may be helpful. This initial placement is dictated by (1) the location of the buildings and open space to be served, (2) the position of the access street, and (3) the relationship the parking lot will have with other site elements, such as trees or a pond.

Once you have determined the location of the parking lot, the buildings, and the open space, lay out the parking lot as a double-loaded corridor arrangement. This will use the least amount of the site area and minimize pavement. Parking layouts often employ drive-through circulation. This means that a driver will enter the site at a specific location along the access street, drive through the parking area, and exit the site along the same or possibly a second driveway to the street. During this process the driver will continue to move in only one direction; there will be no dead-ends and it should never be necessary to drive in reverse.

Parking spaces are drawn using the *Parking Spaces* or *Accessible Space* draw tool. When candidates click on either of these tools, they will see a box that allows them to select the number of continuous spaces desired. Select ten, for example, and you will be able to lay out ten spaces, with the dimensions appearing in the *element information area* as you move the cursor. So, if the required car slot is 9 feet by 18 feet, using this tool to layout a rectangle that is 90 feet in one direction (10 × 9) and 18 feet in the other. At that point, the individual car spaces will automatically appear in the rectangle. Roads are laid out as double lines with the proper width previously programmed. The *draw* tool will allow you to layout the road, with turns or bends in any direction.

Save the placement of trees until all other site elements have been located. Be careful to pay attention to the allowable number of existing trees the program indicates can be removed.

ANALYZING THE SITE PLAN

The solution to a Site Design vignette begins with an analysis of the site plan. A candidate must identify and analyze those conditions that will affect the arrangement of elements on the site. For example:

- If a building entrance is required to receive the noonday summer sun, position that entrance to face due south.

- A service entrance, blocked from pedestrian view, might be necessary for one of the buildings. Position it away from public areas, or place a row of coniferous trees to screen the service drive.

- One of buildings might require a view of a pond. Again, this will affect the arrangement of the elements on the site.

Each site requirement or limitation that can be identified will further define your solution and help determine where key plan elements are placed. Some restrictions will be obvious; a utility easement, for example, will indicate that no construction may occur over the easement area. Similarly, the building limit lines will act as boundaries to your design, beyond which no development may occur. Driveways and pedestrian walks may cross setbacks, of course, but no other development is permitted in setback areas.

Some sites may contain a number of randomly spaced trees. These are placed on the site to create obstacles around which your design must be fitted. In many cases, one is permitted to remove a certain number of existing trees to make the site layout less troublesome. One should not hesitate to remove all the trees allowed, because your solution will come more easily and, in addition, no extra credit is given for removing fewer than the number of trees permitted.

IMPORTANT PARKING OBJECTIVES

There are several ways to run into serious problems in laying out the parking lot in the Site Design vignette. In the interest of avoiding these pitfalls, we offer the following suggestions:

- Never create a dead-end parking arrangement; that is, one in which a driver must back up to reach another part of the parking area or the exit.

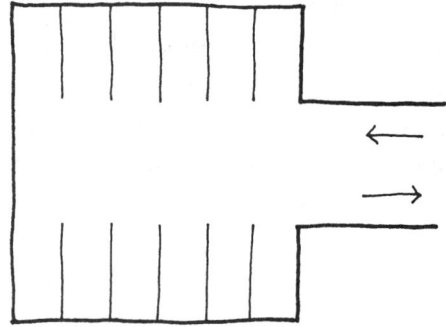

DEAD-END PARKING

- Be certain that the flow of traffic through the parking area is continuous in one direction.
- Always locate the required handicapped parking spaces as close as possible to the entrance of the existing building. It should be obvious that the disabled should walk the shortest distance from their cars to the building.
- Always locate the required handicapped parking spaces so that the disabled are not forced to cross roads or traffic aisles or circulate behind parked cars.

- Always use parking spaces that are perpendicular to (never parallel with) the traffic aisle.

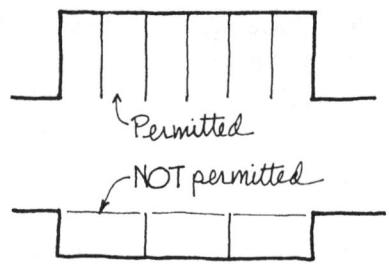

PERPENDICULAR AND PARALLEL PARKING

- Be certain that your access driveway is perpendicular to the existing street for at least the distance specified in the program. The reason for this is safety; a driver should be able to turn with equal ease from either direction when entering or exiting the site. It is not uncommon for an access street to run at an angle to the other sides of the site. Therefore, the rectangular parking area may be parallel to three sides of the site, but the driveway from the street will always be perpendicular to the street for some distance.

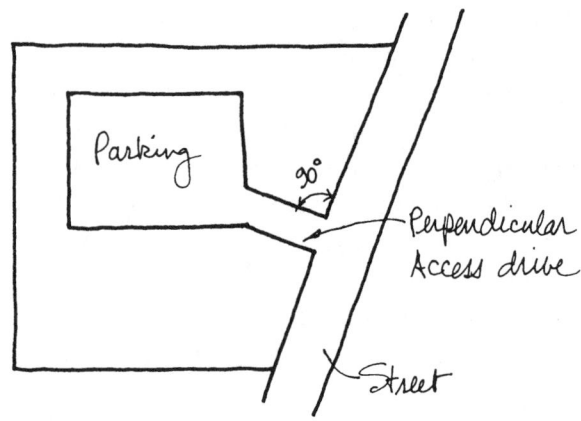

ACCESS DRIVEWAY

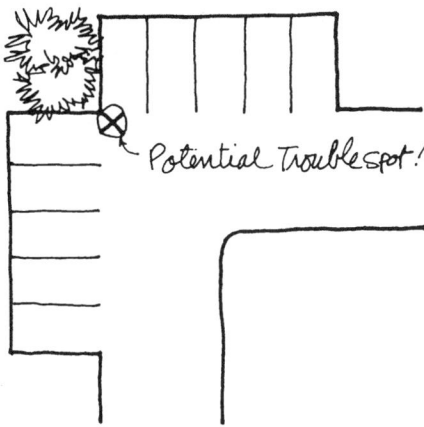

DANGEROUS CORNER PARKING

- A compact and efficient parking layout is highly desirable. However, at right-angled corners, one should never place parking spaces so close to one another that either car could cause an accident when leaving the space.

- If a passenger drop-off area is required, be certain the passenger side of the car ends up adjacent to the drop-off area.

- When avoiding existing trees, be certain that no paved surface encroaches within the drip line of any tree.

After considering the potential pitfalls, some of you may begin to think that parking lot layout is a minefield. However, much of what is listed above is common sense.

ARRANGING THE ELEMENTS

The elements comprising this vignette are listed under the *draw* tool, and clicking on any listed term will allow you to draw that required element. For example, if you click on the term *Classroom Building*, you will be able to drag the prescribed building shape, complete with title, height of structure, and entrance indicated, to any location on the site. You will have no control over the shape of this element; it is predetermined and all you can do is locate it on the site. If, after locating this element, you want to change its position, you may do so with the *move, adjust* tool or with the *rotate* tool. In other words, every required structure may be quickly placed on the site and then arranged later in a final composition.

The elements of this vignette are listed under the *draw* tool in a specific order, and it is probably best to consider each element in that same order. That means the structures will be located first, then the paved plaza or open space, then parking spaces, followed by the driveway(s) and pedestrian walkways, and finally the trees. The position of any element may be adjusted or rotated at any time, so the initial placement is not critical. Nevertheless, a candidate will save a great deal of time if he or she has some notion of how the elements relate to one another, as well as an understanding of all the programmatic restrictions.

CIRCULATION

An essential element of the Site Design vignette is circulation; pedestrians and vehicles must be able to get where they are going safely and efficiently. Most problems require a continuous pedestrian walkway system that includes a paved plaza area off which are the building entrances. The vehicular circulation system begins with access from an adjacent street. The driveway should be perpendicular to the street, and it should lead directly to (1) any or all of the building entrances, (2) the parking area, and (3) the required service entrance. The service drive must generally be kept separate from the parking area. Parking areas should be designed for one-way traffic, as far as possible, and pedestrians should be able to circulate safely from the parking area to the buildings.

FINAL ARRANGEMENT

When you are reasonably satisfied with your schematic arrangement, perhaps 50 minutes or so after beginning this vignette, you must go back to the program and verify every restriction. Some items to check:

- Do the buildings have the required adjacencies and view?
- Are all the components on the site the required distance apart from each other?
- Does the open plaza allow pedestrians access to each of the structures?
- Have the building entrances been shielded from the prevailing winds?
- Has the service entrance—if called for—been blocked from view?
- Have the required number of accessible spaces been met?
- Have all parking lot pitfalls been avoided?
- Have more trees been removed than allowed?
- Do any of the site elements encroach beyond the setbacks?
- Are the easements respected?

Candidates must answer all these questions and verify that there are no encroachments beyond setback lines or easements. To achieve a passing grade, one must simply follow every requirement of the program and solve the problem directly and simply. The computer will never know if your solution is exceptional or mediocre; it will only know whether or not you have violated some restriction.

KEY COMPUTER TOOLS

- **Draw Tool** All the key components of the site plan are included in the draw tool's submenu.
- **Move/Adjust** The candidate will find it absolutely necessary to use this tool for moving components around on the site while trying to arrive at their most advantageous location.
- **Move Group** At times it might be helpful to move multiple elements at the same time.
- **Rotate** An essential tool for positioning the requisite items on the site.
- **Sketch Tool** Use this tool to lay out parking spaces and distances from adjacent elements on site.
- **Zoom Tool** The Zoom tool will be especially helpful for locating the parking lot and driveway accurately on site.

VIGNETTE 1 SITE DESIGN

Introduction

The following Site Design practice vignette includes both general site planning components found on previous Site Design vignettes and a parking lot exercise that is very similar to the stand-alone Site Parking vignettes that were part of previous versions of the ARE. Included in this lesson's narrative is a thorough discussion of all the components of site design: building location, pedestrian open space, parking lot, and vegetation requirements. This study exercise is similar in scope, specifics, and the degree of difficulty that the candidate will encounter on the actual computer examination. The candidate is encouraged to read through this introductory material, do the practice problem, then read the detailed discussion that follows describing the design process and suggested solutions. This will help you gain a better understanding of the recommended approach and design sequencing.

Candidates often wonder where and how to begin a vignette solution. With so little time available, we believe that knowing where to start and what procedures to follow is essential to passing this examination. Without a plan or method, there is little hope of success. Those who are prepared will solve the problem in a direct and logical way, while those using a hit-or-miss approach will waste a good deal of time trying to decide what to do next.

We urge candidates to study the following design procedures. Our solution is certainly not the only possible one, but it evolved through a logical process. If you understand this process, you should be able to apply these concepts to any similar problem.

The Exam Sheet

Shown are our Site Design program and base plan, which were printed on the first sheet of our Site Planning Mock Exam. These simulate the Site Design computer screens candidates will see on the actual exam. The site plan is the same one on which a candidate was required to present his or her solution. Our problems were originally presented on 12" × 18" tracing paper sheets, and most were drawn at a scale of 1" = 40'. The vignettes in this lesson, however, have been reduced to fit the course format, and therefore their scale is 1" = 50'. On the actual exam, scale is relative and generally unimportant. In fact, the appearance of elements may vary with the size of the computer screen.

The program presents relatively simple requirements for a new Church Complex. There are three principal components organized around a Courtyard, but since the Church and Fellowship Hall are combined in a single structure, there are really only two individual buildings to consider. Besides the buildings and Courtyard, the program includes a Parking Area and circulation elements, such as a driveway and pedestrian walks.

Below the printed program are shown graphic representations of the required components of this problem. The two structures must be used as drawn; however, they may be rotated into another position. Also shown are a Fountain, which will be placed somewhere in the Courtyard, and a typical tree to be used to block the wind or an unsightly view. The only important element not shown is the Courtyard, whose area is specified as 10,000 square feet, which is about the size of the Church/Fellowship Hall structure.

VIGNETTE 1 SITE DESIGN

You are to develop a schematic site plan for a new church based on the requirements that follow. The program includes three principal components organized around a central courtyard: worship space, education space, and fellowship space. The required components shall be used as drawn, but they may be rotated.

1. In addition to the required components, provide a Courtyard that is 10,000 square feet in size and which provides direct access to the two structures. Locate the Fountain in the Courtyard.
   - The Courtyard shall be protected from prevailing winds by means of buildings and/or trees.
   - Provide a 40-foot-long minimum passenger drop-off at the Courtyard.
   - Provide pedestrian access to the Courtyard from the street.

2. Locate the School no closer to the Church than 25 feet.
   - The School shall have a view of the existing stream.

3. All driveways shall be 25 feet wide, and circulation shall be essentially one-way.
   - A maximum of two curb cuts are permitted.
   - Indicate the pattern of all vehicular circulation.
   - The service drive may be 15 feet wide, and no turnaround is necessary.
   - The service access shall be hidden from view, as far as possible.

4. Provide parking for 25 cars parked at 90 degrees, as follows:
   - 22 - 10-foot × 20-foot standard spaces.
   - 3 - 15-foot × 20-foot universally-accessible spaces.
   - All parking shall be within building limit lines, and dead-end parking shall not be allowed.
   - The parking area shall be screened from both the Courtyard and Deacon Drive.

5. Additional requirements:
   - The church entrance should be visible from Deacon Drive.
   - No improvements may occur within setbacks or in conflict with existing site features, except driveways and walks may cross setbacks to connect to public access.

Lesson Six: Site Design Vignette

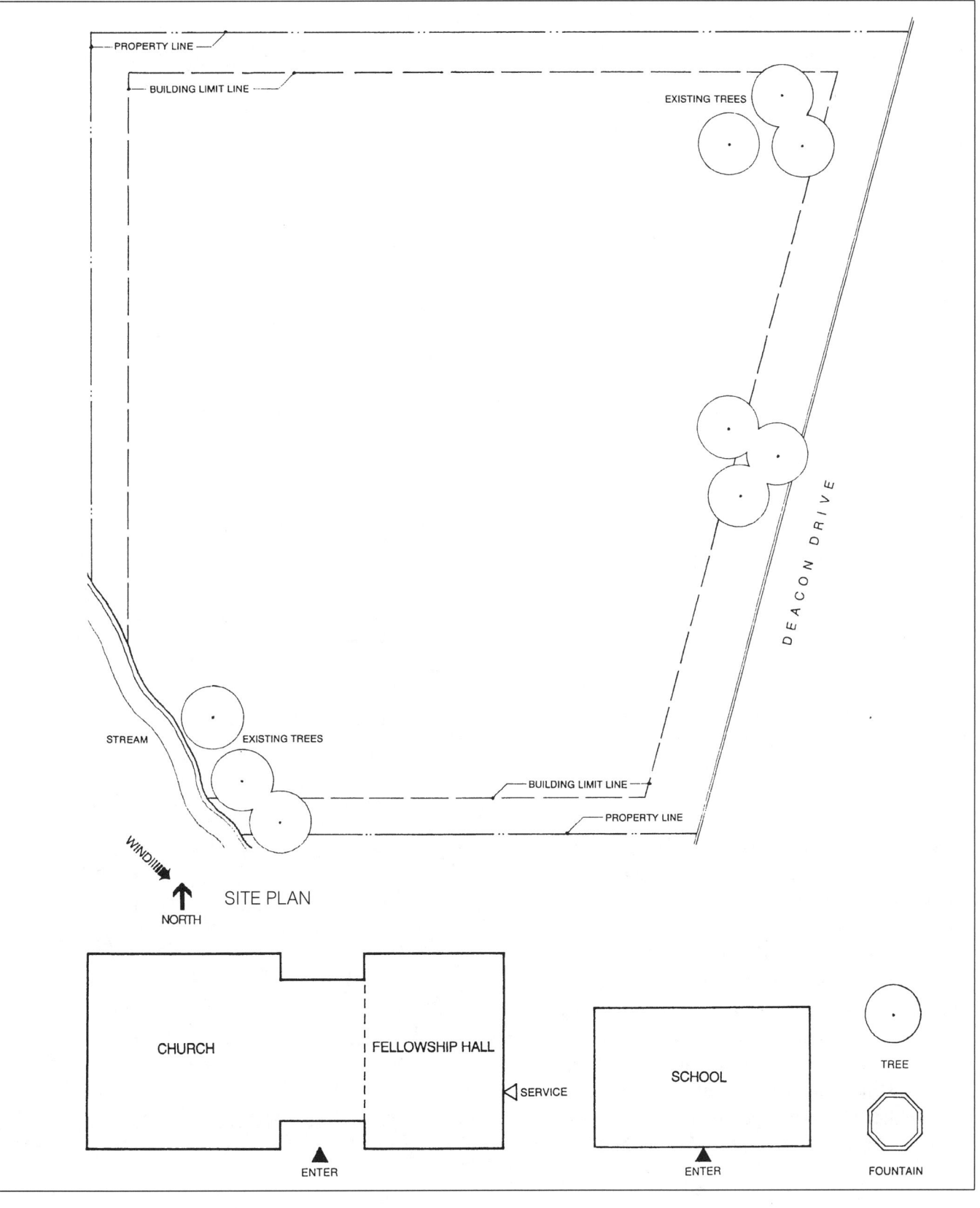

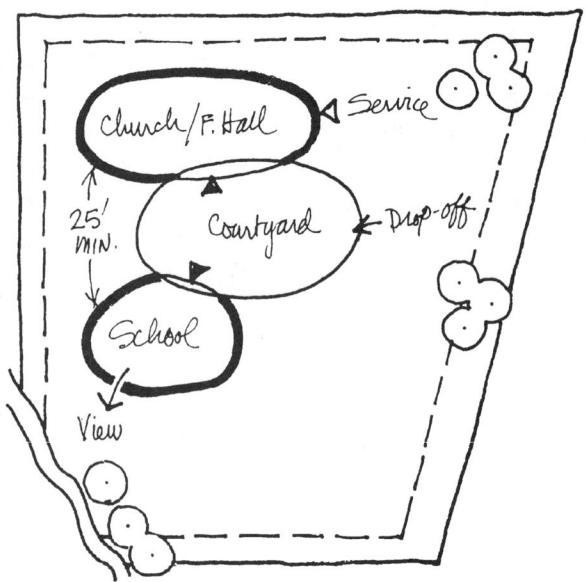

BUBBLE DIAGRAM OF BUILDING LAYOUT

Overall, it is a simple program, but on the other hand, you have only one hour to produce an organized design.

Site Analysis

Candidates should analyze the site to identify those elements that affect the placement of elements on the site. We first notice that there is a single access street, Deacon Drive, from which all pedestrian and vehicular traffic will come. This will have a significant impact on the location of the access drive and parking lot. The program states that the School must have a view of a stream, and we note that this stream is at the southwest corner of the site. There are a few existing trees on the site, and these are placed here as hazards, similar to sand traps on a golf course. One must work around them, since they must not be encroached upon or removed. Building limit lines are shown, and all development must be contained within these. Finally, we note the north arrow pointing straight up and the wind direction pointing to the southeast.

Bubble Diagram

It is generally best to locate the largest structure first, in this case, the Church/Fellowship Hall. The only restrictions are that it must face the Courtyard and that its entrance must be visible from Deacon Drive. We assume the Courtyard will be near Deacon Drive, because pedestrian access is required to the Courtyard from the street. Therefore, we begin by locating the Church/Fellowship Hall at the northwest corner of the site. We are not yet certain of the building's orientation. The entrance may face south or east, which means that the service entrance will face east or north. At this location, we have sufficient room to place the School toward the southwest corner, where it can overlook the stream.

We next locate the School south of the Church/Fellowship Hall, where it will be close enough to the stream to have the required view.

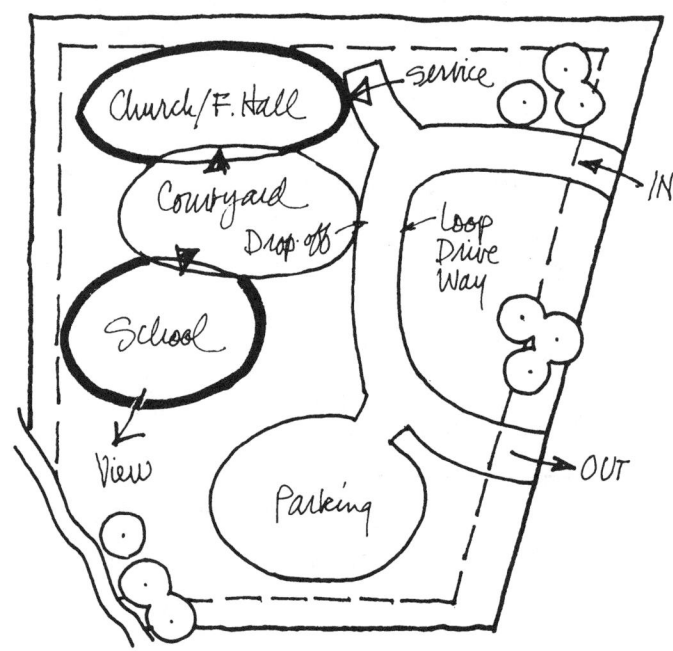

BUBBLE DIAGRAM WITH CIRCULATION

However, it must be at least 25 feet from the Church/Fellowship Hall. The School entrance must face north or east, so that it can be accessed directly from the Courtyard. We can now lay out the 10,000-square-foot Courtyard, which incorporates the entrances of both buildings. Regarding this element, or any paved open space you are given, always assume it will be a rectangle whose proportions are between 1:1 and 2:1.

Circulation

Our next concern is the vehicular circulation on the site. The program calls for one-way traffic, and two curb cuts are permitted. From these two facts, one should immediately visualize cars entering the site at either the north or south end of Deacon Drive and exiting at the opposite end. How do we decide which end will be the entrance and which the exit? The answer lies with the requirement for a 40-foot-long passenger drop-off at the Courtyard. To position the passenger side of the car along the Courtyard, cars must enter the site at the north and exit at the south.

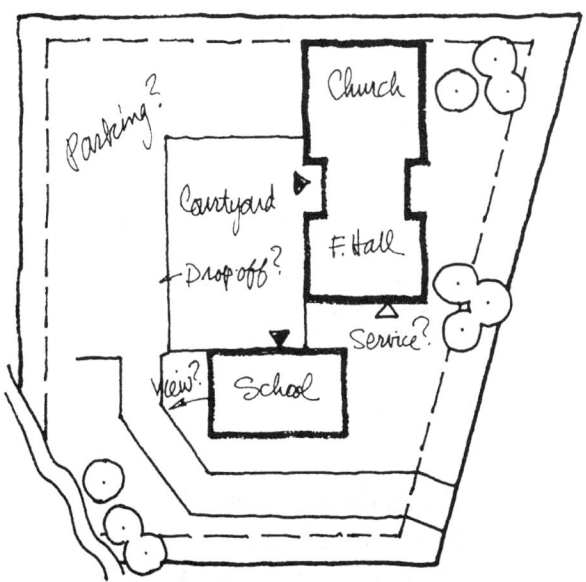

UNWORKABLE SOLUTION

Thus, we create a driveway loop off of Deacon Drive that is adjacent to the Courtyard. These are the only two curb cuts permitted; therefore, the Parking Area and service drive must somehow connect to this driveway loop. Since the Service Drive must attach to the Fellowship Hall, we assume that the Service Drive will head northward, while the Parking Area will be located at the southern end of the site.

Diagrammatic Arrangements

Much of the site analysis discussed above may be no more than a mental process, and candidates may not have the time or the need to draw the assumed location of every element. It will generally be sufficient to use common sense in eliminating most possibilities. For example, if one considered locating the structures along Deacon Drive, it would be immediately apparent that the Courtyard would have to be west of the buildings, and thus, the building entrances would be out of sight from the street. In addition, the Courtyard would be completely unprotected from the prevailing winds, and the vehicular circulation and parking would be difficult to arrange. You do not have to draw solutions such as these, because it should be obvious that they do not work.

With the position of the major elements determined, we can now proceed to design the site. As shown in the 1st Diagrammatic sketch, we locate the Church/Fellowship Hall at the northwest corner of the site, and the School is placed 25 feet south of that structure, close to the stream. The building entrances are 90 degrees to one another and connected by the rectangular Courtyard. The position of the two structures will protect the Courtyard from the prevailing winds. We then locate the access driveway adjacent to the Courtyard to provide for the passenger drop-off. The northern end of the driveway turns into a Service Drive which runs to the Fellowship Hall. The Parking Area in this

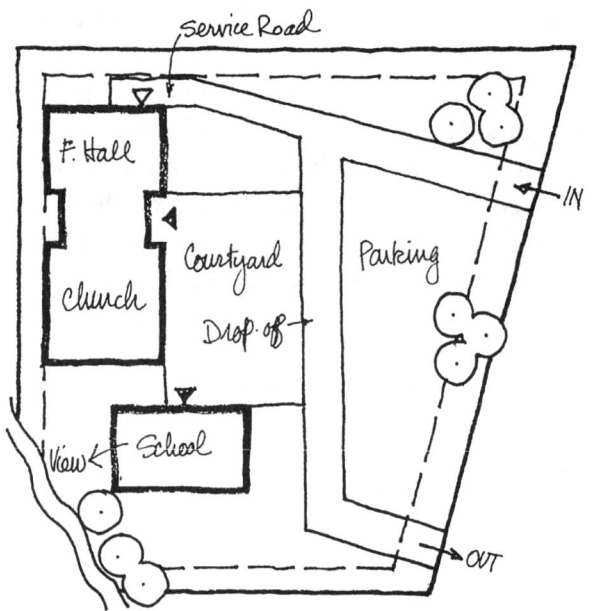

1ST DIAGRAMMATIC SCHEME

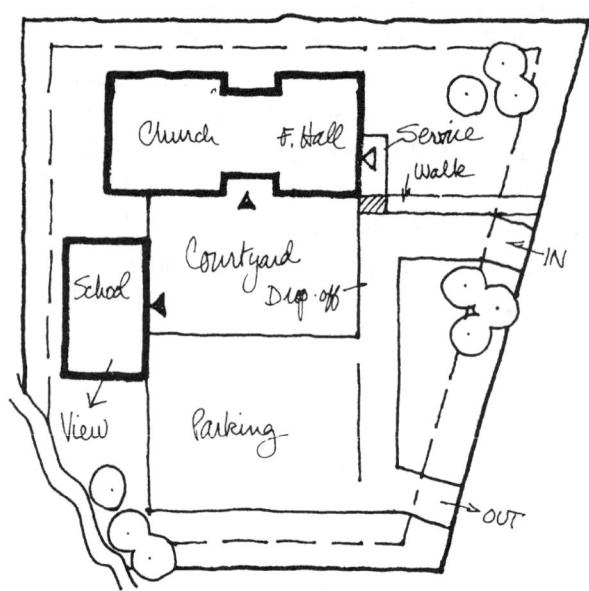

3RD DIAGRAMMATIC SCHEME

scheme is placed along Deacon Drive, across the driveway from the Courtyard.

This 1st Diagrammatic solves most of the problems, but a remaining problem is the location of the Parking. First, the view of cars is all one sees from Deacon Drive. More serious, however, is that all drivers must cross the driveway to reach the buildings. This kind of circulation is unsafe and undesirable.

The 2nd Diagrammatic sketch shows the Church/Fellowship Hall at the same northwest corner, but rotated 90 degrees. The Courtyard separates the two structures, and the School's view of the stream is maintained. The loop driveway provides access to the Service Road, Courtyard drop-off, and Parking Area, which has moved to the southern part of the site. In all, this arrangement works fairly well; however, the Courtyard is exposed to the prevailing winds and the Parking Area is just a bit too small.

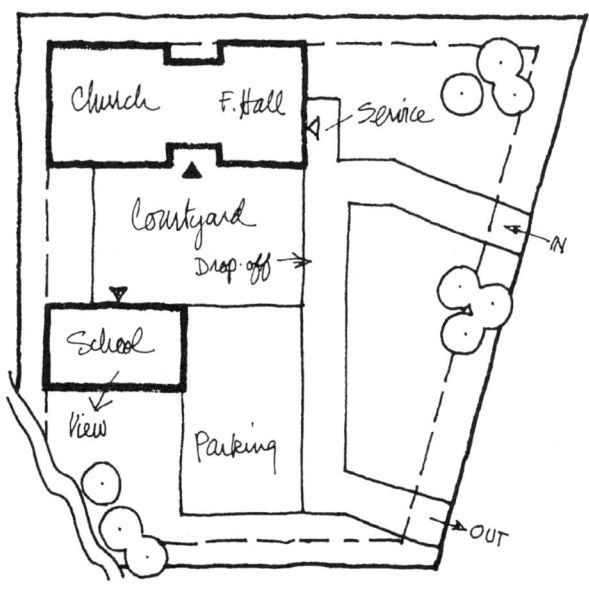

2ND DIAGRAMMATIC SCHEME

Our final Diagrammatic sketch is a refinement of the previous arrangement. The Church/Fellowship Hall has moved to the east; the School is located 25 feet to the south and rotated 90 degrees. The Courtyard

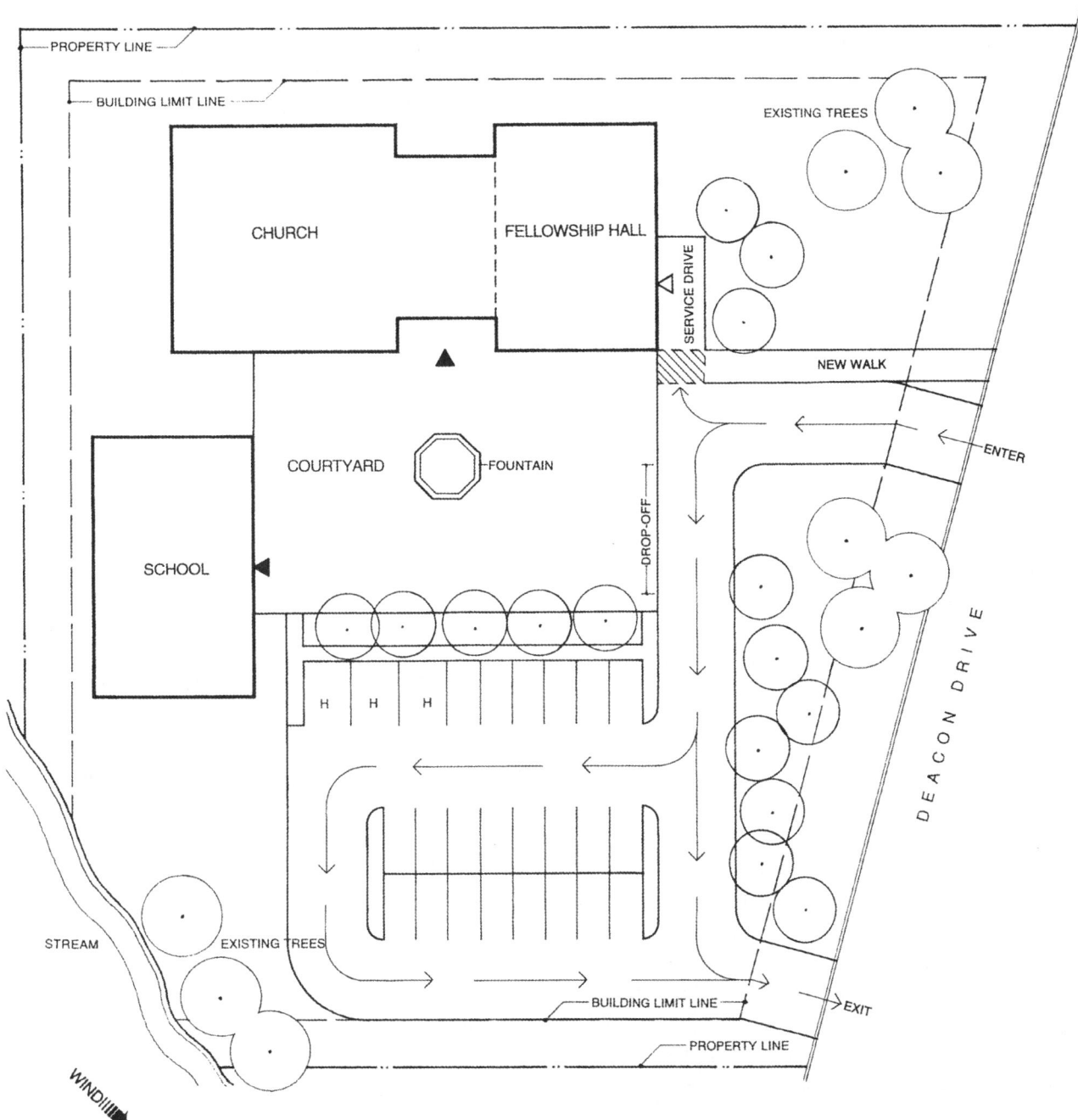

SITE DESIGN VIGNETTE - SUGGESTED SOLUTION

accommodates the entrances of the two buildings, and the Parking Area is now larger and located directly south of the Courtyard. The looped driveway remains, and we have added the required pedestrian walk. Vehicles now enter the site at about the middle of Deacon Drive, turn right for the service drive or left for the 40-foot-long passenger drop-off, and then continue south to the Parking Area. This layout also allows vehicles to exit the property at the south end of Deacon Drive without having to drive through the parked cars.

Potential Problems

At this point, we should review the program to be certain all the elements are there, and that we have not created any insurmountable problems. We note that the service drive, which is required to be hidden from view, faces directly onto Deacon Drive. This will probably have to be screened by trees, because to rotate the building and face the service access northward would cause even more serious problems.

Another potential problem is locating the pedestrian access from the street to the Courtyard. Regardless of where we place the walk, pedestrians must cross the driveway to reach the Courtyard. This could create a hazardous situation, and therefore, the problem must be solved when we make our final layout.

Final Site Design

Our final Site Design is shown superimposed over the site plan on the first preprinted sheet. The elements are shown to scale, and all the other requirements have been completed. For example, the Parking Area is shown with the required 25 cars, including the three handicapped spaces located as close as possible to the Courtyard. The Parking Area was to be shielded from both the Courtyard and Deacon Drive, and this is accomplished by adding two screens of trees.

Vehicular circulation is shown as previously planned, and traffic is indicated by means of directional arrows. It is essentially one-way traffic that begins at Deacon Drive, passes by the passenger drop-off, and then continues to the Parking Area or to the exit at the southern end of Deacon Drive. The new pedestrian walk from the street was placed just north of the entrance driveway. Where it crosses the narrow service road, we have indicated hatching, which recognizes our concern for this less than perfect situation. Our reasoning here is that the service drive traffic will be less frequent and busy than that of the main driveway. It is always best to keep pedestrians and vehicles widely separated, but with this arrangement, that was not possible.

Finally, we placed the required Fountain near the center of the Courtyard, on the axis of the Church/Fellowship Hall entrance. The shape of the Fountain indicates that this is an element without a front or back and probably equally interesting from all sides. Thus, it should be placed where pedestrians can circulate around it.

Before moving on to the next problem, it is a good idea to return one last time to the program to verify all required elements, relationships, and restrictions. This should take no longer than a few moments, and the time invested may be considered additional insurance against an unintentional oversight.

LESSON SEVEN

SITE ZONING VIGNETTE

Introduction
Vignette Information
Design Procedure
Key Computer Tools
Vignette 2 Site Zoning
  Introduction
  The Exam Sheet
  Drawing the Secondary Construction Area
  Drawing the Buildable Area
  Drawing the Grade Profile
  Drawing the Building Envelope
  Conclusion

INTRODUCTION

The second vignette in the Site Planning division now combines two previous vignettes, Site Zoning and Site Analysis, into a single 1-hour Site Zoning vignette. This problem evaluates a candidate's understanding of two important site-related issues:

1. Cross-sectional building area limitations imposed by zoning and other setback restrictions (site zoning)

2. Factors that influence the subdivision of land and determining suitable areas on a site for construction of buildings and other surface improvements, such as parking areas (site analysis)

In this vignette the you are given a program, a site plan, and a cross-sectional grid on which the zoning solution is to be drawn. The site plan will indicate existing lot lines, streets, and other features. This parcel will need to be divided into two lots and labeled as to the buildable areas for structures and other site improvements, based on the programmatic requirements. On the zoning grid, you are to draw a schematic section that includes (1) the existing site grade and (2) the maximum building envelope above that grade allowed by the program's restrictions.

VIGNETTE INFORMATION

As before, the Site Zoning vignette begins with an index screen, which lists other information screens, such as the following:

- **Vignette Directions**—This screen describes the site plan and the grid in general terms. You are directed to accomplish four tasks:

  1. Outline the area on the site plan suitable for the construction of surface improvements

  2. Outline the area on the site plan suitable for the construction of buildings only

  3. Draw the profile of the existing grade line on the grid below the site plan

  4. Draw the profile of the maximum building envelope, in accordance with program and site conditions

- **Program**—This screen begins with a general description of the site plan and what you are expected to accomplish in the exercise. The screen continues with a detailed description of restrictions that will dictate the parameters of the design, including:

    1. Location of property lines
    2. Setback information for surface improvements
    3. Setback information for the construction of buildings
    4. Easements
    5. Height limits
    6. Solar access planes
    7. Other site conditions

- **Tips**—These are suggestions about procedures intended to make a candidate more efficient, such as using the *sketch* tools, and particularly the *line* tool, to plot the elevations of the section profile. Other helpful tools include the *full screen cursor* and the *ortho* tool, which insures that a line projected from a point on the plan to the grid will be perfectly vertical.

- **General Test Directions**—These are the same directions that apply to all vignettes, and they may be reviewed at any time for any vignette. However, one reading is probably all that is necessary for most candidates.

The preprinted site plan and sectional grid on which candidates are required to present their solution are found on the work screen. The site plan generally includes property lines, an access street, topographic contours, a bench mark elevation, and a cross-sectional line that runs horizontally across the middle of the site plan. The cross-sectional grid is graduated both vertically and horizontally so that one can visually measure height and length. Again, one can find the computer tool icons displayed along the left side of the screen.

DESIGN PROCEDURE

Of the three vignettes in the Site Planning division, this is the most straightforward and mechanical, requiring not so much design skill as close attention to the restrictions and limitations and drawing the profiles and buildable areas as required. For each part of the exercise there is only one solution possible:

- One area on the site plan suitable for the construction of surface improvements
- One area on the site plan suitable for the construction of buildings only
- One profile for the existing grade line on the grid
- One profile for the maximum building envelope on the grid

Begin the site plan buildable areas by determining the setbacks for surface improvements and the construction of buildings. Determine the maximum area allowable for the construction of surface improvements first. Use the *sketch* tool or *sketch grid* tool to lay out the limits of the secondary construction area. Under the *draw* tool on the site plan screen you will find the necessary tool to outline the boundaries of the secondary construction area. Once you have established the secondary construction area, lay out the buildable area in the same fashion.

To draw the profile of the grade line and building envelope on the grid, project the elevations along the section cut line on the site plan to the cross-sectional grid.

KEY COMPUTER TOOLS

- **Sketch Tool** On the site plan, use this tool to measure the distances for setbacks. On the cross-sectional grid, this will be an indispensable tool for helping layout the elevations and building profile before using the draw tool to add the grade and buildable area profile.

- **Ortho Tool** Candidates will most likely find this tool of great assistance in laying out the site. On the section, this tool will be helpful for keeping elements in line with each other, the site plan, and the grid.

- **Full Screen Cursor** Similar to the ortho tool for working on the cross-sectional grid, the Full Screen Cursor aids in keeping components lined up.

VIGNETTE 2 SITE ZONING

Introduction

The following Site Zoning vignette asks you to determine buildable areas on the site plan and to draw the existing grade and buildable profile on the cross-sectional grid. As with the other vignette examples, our solution is presented in a logical progression of steps in which each element of the problem is considered in sequence. Those who learn to solve vignette problems this way should have few difficulties.

The Exam Sheet

Shown are the Site Zoning program, site plan, and site section grid on which candidates were to present their solutions. You may notice that the plan and section grid have no scale, because all the dimensions one needs are indicated on the grid.

The program directs you to draw the following:

1. An outline of the area on the site plan suitable for the construction of surface improvements
2. An outline of the area on the site plan suitable for the construction of buildings only
3. The profile of the existing grade line on the grid
4. The profile of the maximum building envelope on the grid

The profile of the existing grade and the allowable building envelope will be drawn along the section cut line X-X shown on the site plan.

The site plan shows a narrow plot of land that runs basically east and west. At the front of the property is Sun Street, and at the center of the curb is noted a bench mark elevation of 50 feet. Along the length of the property we note randomly spaced contours, which are essentially vertical. Our only concern about these contours is their elevation at the points where they cross the section cut line noted as X-X. On this plot the candidate is to determine the limits for the construction of surface improvements and limits for the construction of buildings. A utility easement is indicated between the two lots.

The site section grid is a simple graph with horizontal and vertical dimensions shown along the bottom and right side. Although the horizontal and vertical dimensions are at the same scale, that is, one small square equals 10 feet in both length and height, the two scales could conceivably be different. It should matter little to you, since you are drawing an abstract diagram.

Drawing the Secondary Construction Area

The Site Zoning vignette requires the examinee to identify the secondary construction limits of the site—the area where surface improvements can be made, excluding buildings. The program defines the parameters for the secondary construction limits. The logical thing to do is draw these setbacks on the lots and infill the resulting space with a hatch and label it "Secondary Construction." The secondary construction area will in all likelihood be more generous in size that the buildable area on site. Conceivably, the secondary construction area might match the buildable area. At no time, however, may the buildable area be larger than the secondary construction area.

In this vignette the limits of the secondary construction zone is fairly straightforward to determine. It is simply 5 feet in from every property line, excluding the utility easement, as shown on the next page.

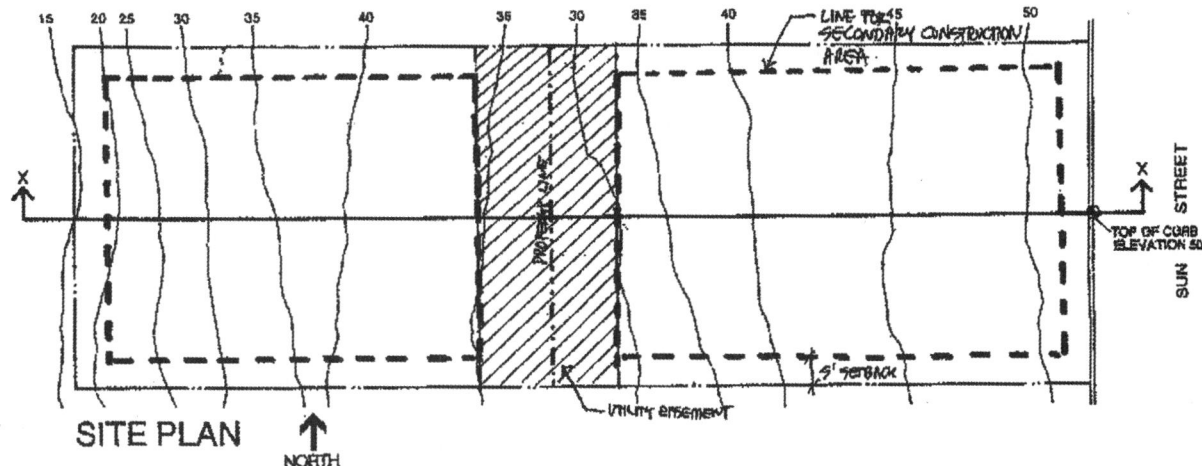

Drawing the Buildable Area

Once the secondary construction area has been established, it is easy to layout the area in which buildings can be constructed, or the "buildable area." The program will inform you of the setback requirements and other stipulations for determining this area. For this exercise, we discover that the setback from Sun Street is 15 feet. The candidate simply draws a line parallel 15 feet from Sun Street. The side-yard setbacks for the buildable area match the limits of the secondary construction zone, 5 feet from the side property lines. This will not always be the case. The buildable area will often be more restrictive than the area allowed for surface improvements. Finally, we see that the rear setback is 10 feet from the property line. One thing to note in this problem is that there is a special condition for Lot A. The program informs us that the front and rear setback for this property is 10 feet. The candidate will also note that no building of any sort is allowed in the utility easement.

We now have the limits for buildable area. The examinee may wish to use a different hatch to indicate this area to differentiate it from the secondary construction area.

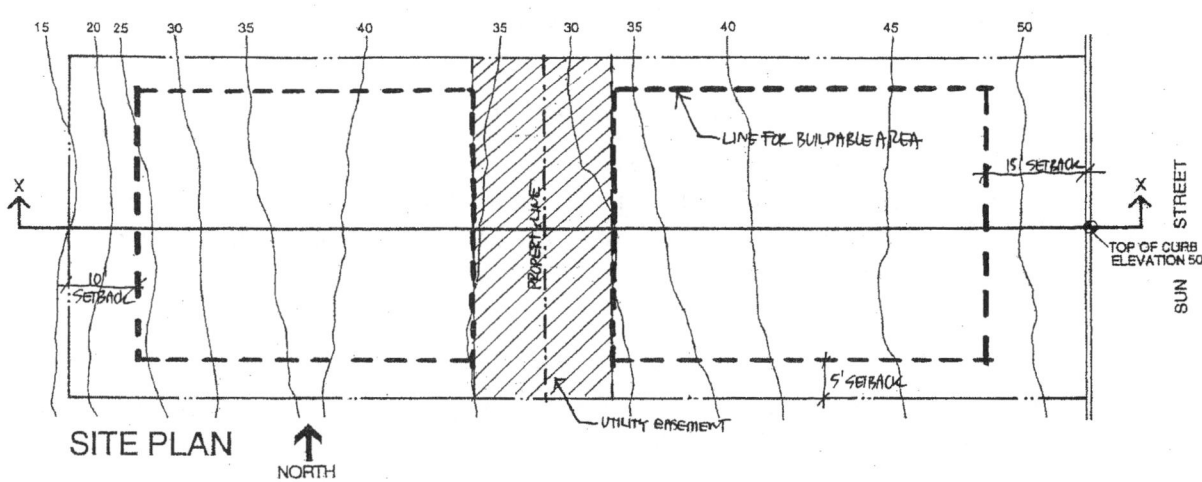

VIGNETTE 2 SITE ZONING

Using the site plan and cross-sectional grid drawings you are to accomplish the following tasks: on the site plan, identify the secondary construction area and the area for the construction of buildings on the two lots shown. On the cross-sectional grid, two profile lines must be drawn: one indicating the existing grade and the other outlining the maximum building envelope. All restrictions and regulatory requirements are to be met.

1. On the plan, show the portion of the each lot where surface improvements are allowed.
2. On the plan, show the portion of the each lot where building construction is permitted.
3. On the grid, draw the profile of the existing grade at section X-X.
4. On the grid, draw the profile of the maximum building envelope for each lot at section X-X.

Observe all the following restrictions:

- Surface improvements are prohibited within 5 feet of any property line.
- Construction of building or surface improvements are prohibited within the utility easement.
- Construction of buildings is prohibited within the following setbacks:
    - Setbacks from Sun Street are 15 feet.
    - Rear yard setbacks are 10 feet.
    - Side yard setbacks are 5 feet.
    - Front and rear setbacks on Lot A are 10 feet.
- Maximum height limit within 75 feet of Sun Street is 30 feet above the curb elevation.
- Maximum building height is 45 feet above the curb elevation.
- The maximum building envelope is restricted to an elevation defined by a 45-degree line rising eastward from a point on the rear setback line on Lot A at an elevation equal to the curb elevation and ending at the maximum building height elevation.

Lesson Seven: Site Zoning Vignette

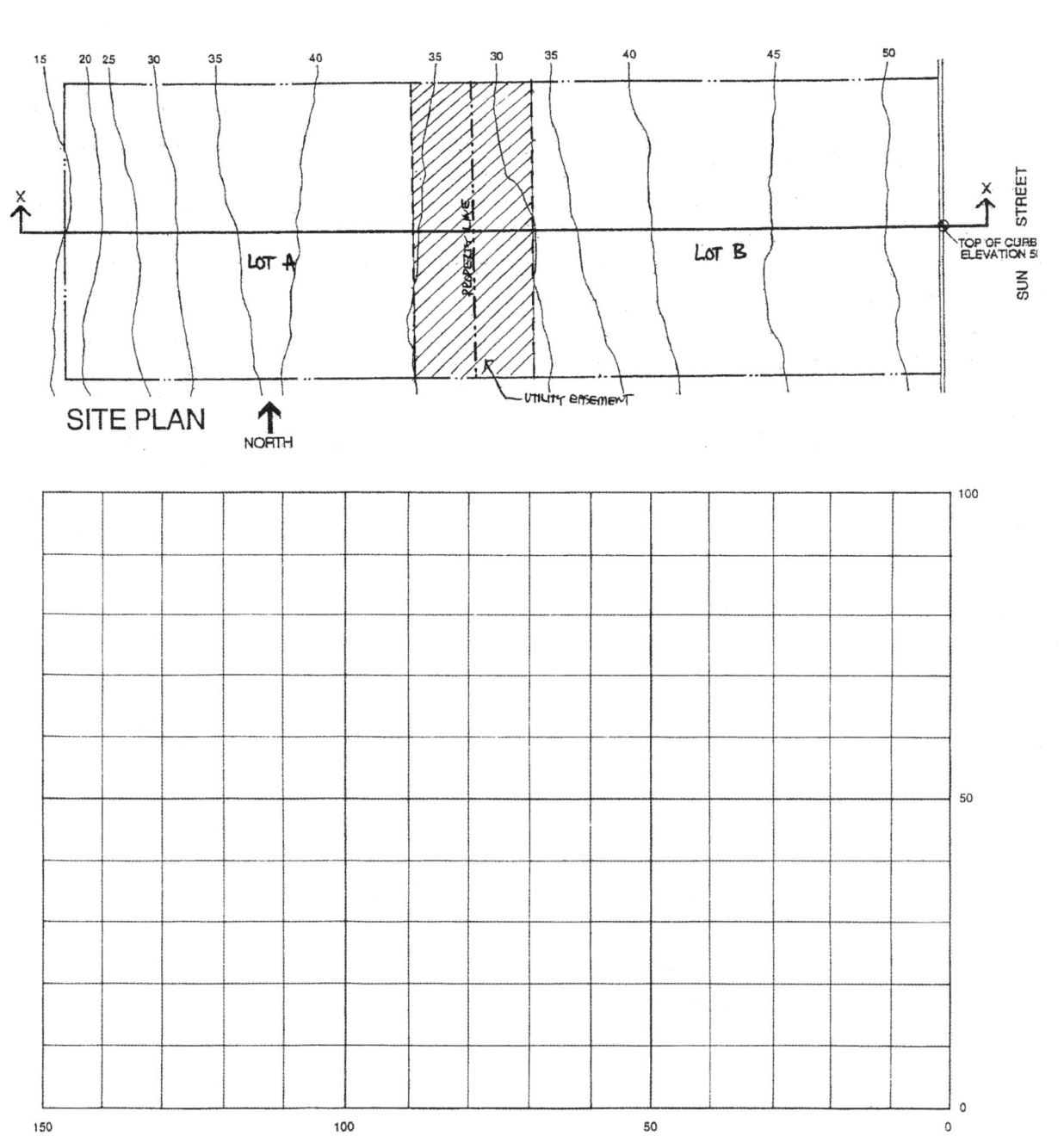

SITE SECTION - VIEW TOWARD THE NORTH

Drawing the Grade Profile

The solution to the sectional profile of the Site Zoning vignette consists of two distinct parts, the existing grade profile and the maximum building envelope. You must always begin with the existing grade profile, and one may plot this profile from either end of the site. If we begin at the Sun Street side of the property, we note that the top of the curb is indicated as elevation 50. Without projecting any lines, we can immediately go to the grid and place a dot at the 50 level, along the far right vertical line.

The following points are determined by noting the intersections of the section cut line with the contours. Candidates should remember that a contour line is simply a line that connects points of equal elevation. Thus, every point along the 50 contour, including where it intersects the cut line, is exactly at elevation 50. When that point is projected downward to the grid, we

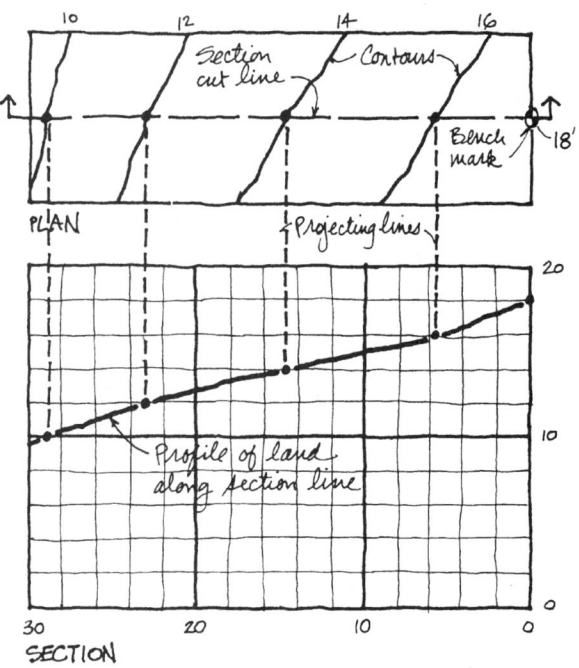

DRAWING A GRADE PROFILE

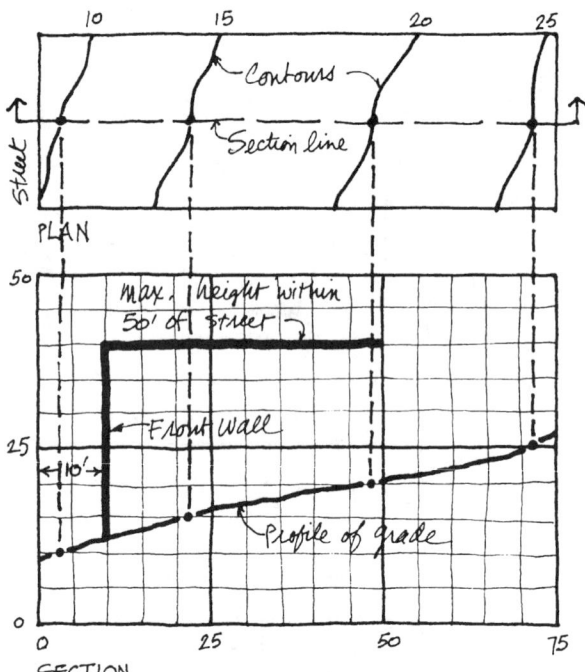

MAXIMUM BUILDING HEIGHT

already know its position relative to the vertical scale; it is at elevation 50. The information we are projecting is its position relative to the horizontal scale, the left-to-right scale at the bottom of the grid.

The second part of the problem is a bit more complicated. Drawing a sectional profile of the maximum buildable envelope will depend on the restrictions found in the program. For example, if the program stated: *the setback from the front property line is 10 feet*, then one would draw the front wall of the building section 10 feet away from that property line. And if *the maximum building elevation within 50 feet of the street is 40 feet* (measured on the graph's vertical scale), then the same front wall would be limited to that specific height.

Incidentally, height limits may be stated in a variety of ways. For example, the height limit may be a specific value, such as: *the maximum building elevation is 80 feet.* The building height may also be expressed as a certain number of feet above the property line, or above the bench mark elevation (wherever that mark is located), or a certain number of feet above the bench mark elevation within a certain distance of a property line or street. In other words, one must be prepared to decipher the convoluted language of this exercise and then follow the directions precisely.

The restrictions of the Site Zoning vignette generally include front and rear yard setbacks, height limitations, and possibly an easement restricting all development within its boundaries. As an example, the program might read as follows: *Construction is prohibited over a 20-foot-wide utility easement whose center line runs parallel to and 60 feet distant from the rear property line.* The delineation of this restriction is illustrated on the next page.

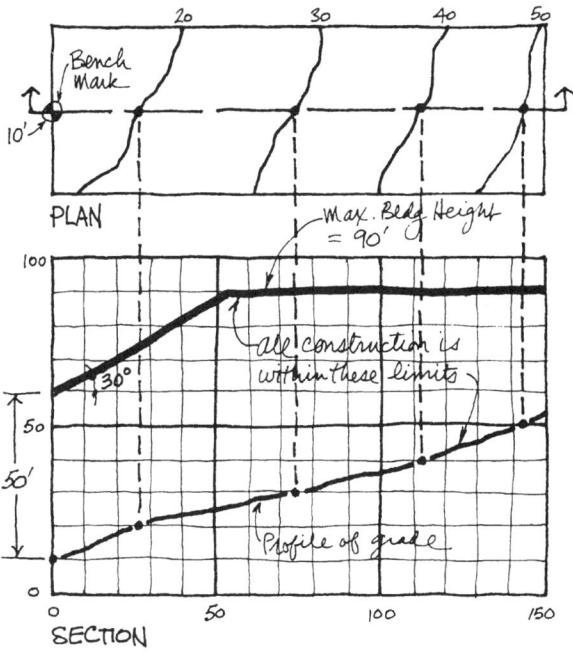

SOLAR ACCESS PLAN

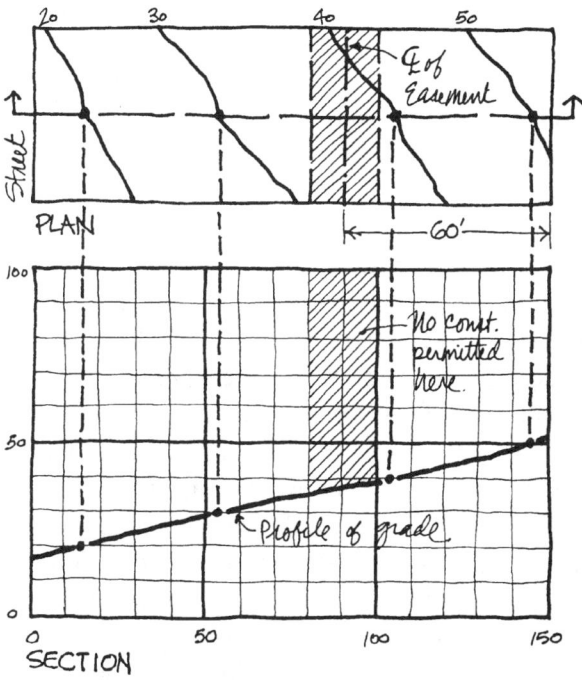

EASEMENT RESTRICTION

There is one more restriction that one may find confusing. It is often known as a *solar access plane*, and an example of this restriction might be stated as follows: *The maximum building envelope is restricted to an elevation defined by a 30-degree line rising eastward from a point on the front property line that is 50 feet above the bench mark elevation and ending at the maximum building height elevation of 90 feet.*

The language of this restriction may take another reading or two and perhaps a diagram such as the one that follows. It is not the design concept that is difficult to understand, it is the expression of the idea.

All programmatic restrictions must be considered and plotted one at a time until they form a complete building profile. You must remember that you are not designing a structure; you are simply following the directions and restrictions of the program until the structure's profile is completed.

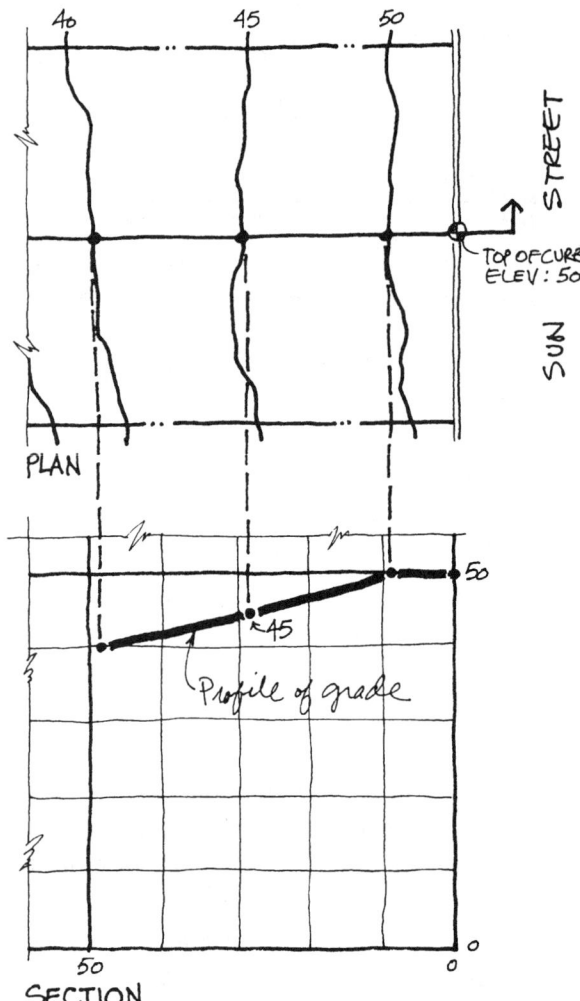

PLOTTING THE GRADE

We place a dot on the grid where indicated, and thus, we see the lot is level for the first 10 feet or so. The next contour to the west is the 45 contour, and again we project its intersection with the section cut line down to the 45 level of the grid, which is located midway between the 40 and 50 designations.

In a like manner, we proceed westward and find the intersections of the remaining contours with the section cut line. We project these points downward and place a dot at the appropriate spots on the grid. When we have completed this mechanical exercise, we will see a series of dots on the grid, which represent the elevations of all the contours cut by the section cut line in plan. If we connect the dots, we will have an accurate profile of the land along the section line. One must be careful not to omit even a single contour intersection, because such an omission will alter the profile and lower your test score.

Drawing the Building Envelope

The term *building envelope* refers to the outline of a structure or the limits of its enclosure. In the case of the Site Zoning vignette, we are referring to the section cut through a structure that will enclose the maximum allowable space.

It is best to begin this process by following the order of the program's several restrictions. The first of these notes the setback from Sun Street as 15 feet. Thus, we can locate a vertical line that is 15 feet west of the zero on the bottom scale. Similarly, we note that the rear yard setback is 10 feet, and so we locate another vertical line 10 feet east of the left edge of the grid, which is at the 140-foot mark on the bottom scale. With these two vertical lines, we have defined the left and right limits of the building.

The third restriction states that the maximum building height within 75 feet of Sun Street is 30 feet above the curb elevation. Since the curb elevation is noted as 50 feet, 30 feet above this level would be 80 feet. Returning to the vertical line at the east, we find the 80-foot level and draw a horizontal line running westward. But how far west should this line be drawn? The requirement says the 30-foot height limit extends 75 feet from Sun Street. However, the utility easement starts 70 feet from Sun Street. Since the east building line is already 15 feet from Sun Street, we draw the horizontal line another 55 feet to the west.

At that point, the next restriction takes precedence: the utility easement. The program indi-

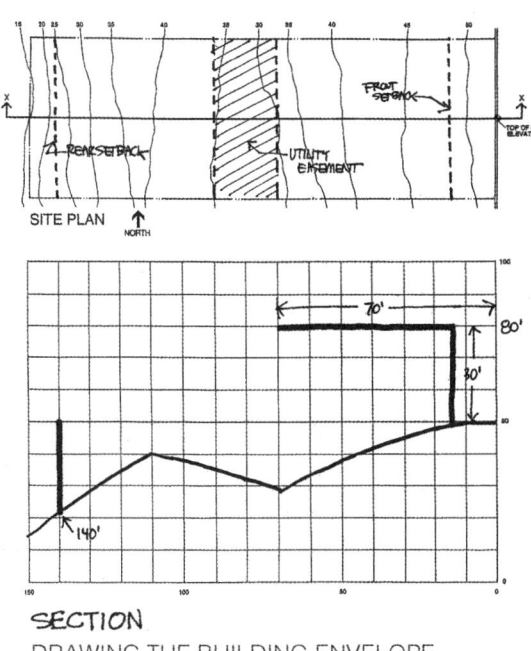

SECTION
DRAWING THE BUILDING ENVELOPE

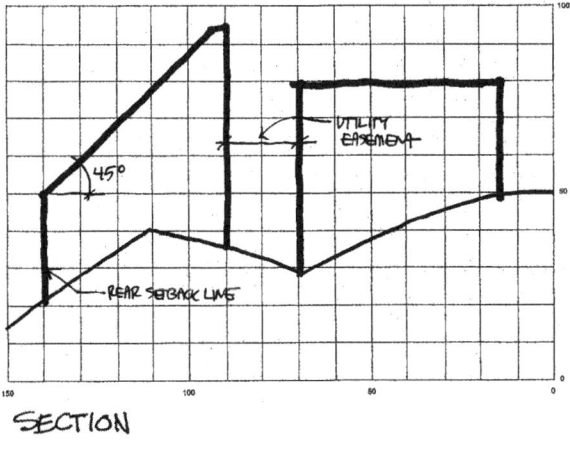

SECTION
COMPLETING THE BUILDING ENVELOPE

cates that no building can occur within the utility easement, so the line will drop down to grade at the 70 foot mark. It will start from grade again at the 90 foot mark, which is the west side of the utility easement. At this point the maximum allowable building height changes. It is 45 feet above the curb elevation here, or 95 feet according to the grid lines. The vertical building line we start at the horizontal 90 foot mark will extend up to the 95 foot elevation. This new roof line will now continue westward.

Our final restriction should allow us to complete the diagram. However, this restriction is the most complicated. It states that *the maximum building envelope is restricted to an elevation defined by a 45-degree line rising eastward from a point on the rear setback line at an elevation equal to the curb elevation and ending at the maximum building height elevation.* To understand this wordy requirement, it is best to take it one step at a time. First, the limit line will be a line angled at 45 degrees. Thus, when the requirement says you are restricted to *an elevation,* it really means that you are restricted to all elevations falling along this 45-degree line.

Next, the 45-degree line originates along the building's rear setback line, which is at the 140-foot mark measured along the lower grid scale. Its beginning elevation equals the curb elevation, which is 50 feet. With this information we can establish a point on the grid along the rear setback line, at an elevation of 50 feet. From this point we project a 45-degree angle eastward until it intersects the maximum building height of 95 feet. The geometric shape is closed, and our solution is complete.

Conclusion

The discussion of this solution is perhaps more lengthy and complicated than the actual problem. Success on the Site Zoning vignette is simply a matter of following every direction precisely. If one complies with every restriction, and avoids simple mistakes, there is no reason why he or she should not succeed on this vignette. Those who fail are those who lose their concentration, or ignore, misunderstand, or misinterpret a requirement.

Site Planning

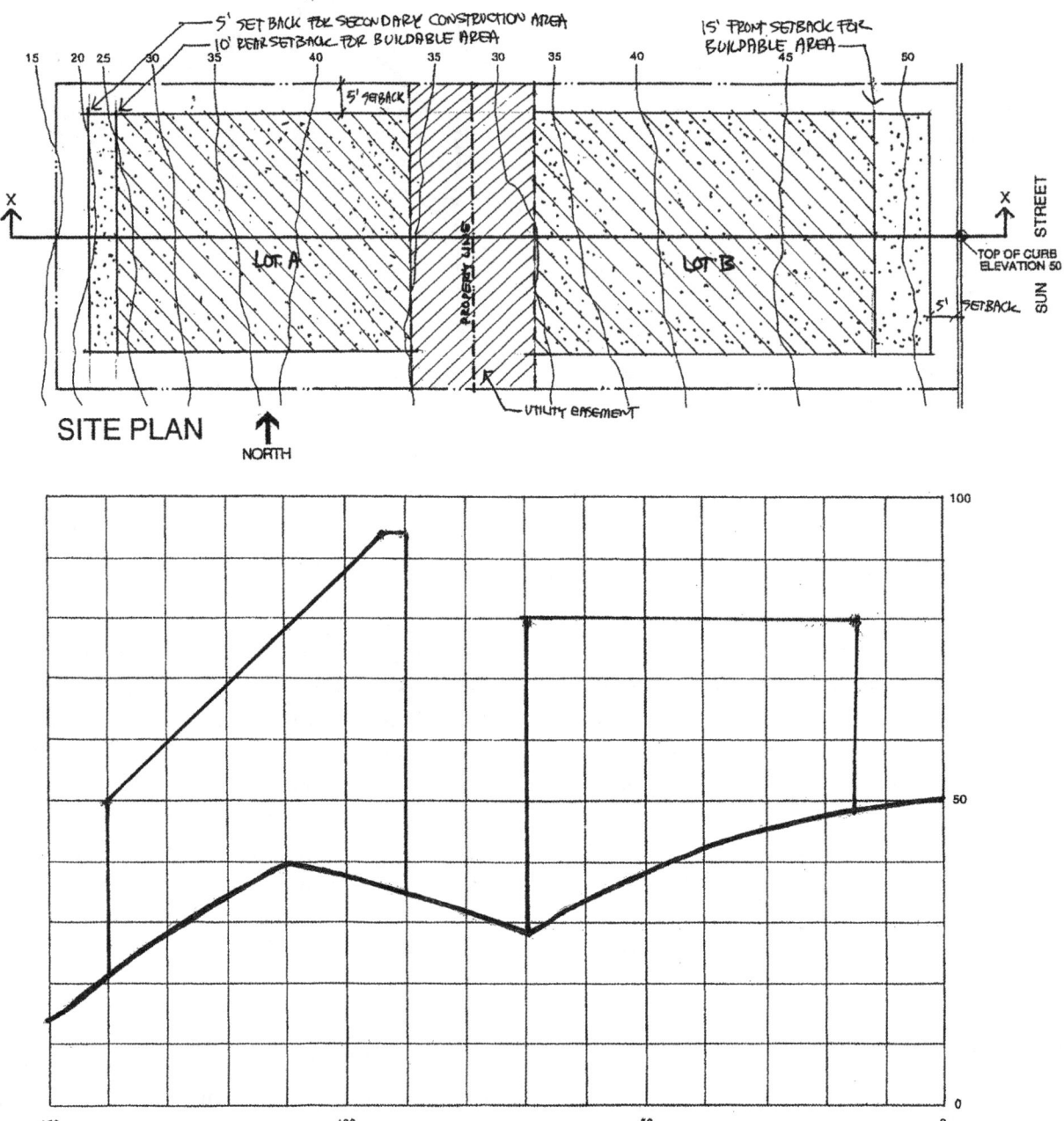

SITE ZONING VIGNETTE - SUGGESTED SOLUTION

LESSON EIGHT

SITE GRADING VIGNETTE

Introduction
Vignette Information
Design Procedure
Key Computer Tools
Vignette 5 Site Grading
Introduction
The Exam Sheet
Regrading the Parking Lot

INTRODUCTION

The final vignette on the Site Planning exam tests a candidate's ability to modify the topography of a site. You are given a program and a site plan that includes existing contours, landscaping features, and man-made elements. You are then asked to modify the existing contours to accommodate the given elements and satisfy the drainage requirements and other restrictions of the program. Solutions are analyzed for compliance with the programmatic requirements, completeness, and technical accuracy.

VIGNETTE INFORMATION

The Site Grading vignette begins with an index screen from which one may access the other information screens as follows:

- **Vignette Directions**—The directions on this screen are brief and quite simple. You are told that the accompanying topographic representation of an existing site is to be regraded. You are then directed to modify the contours so that water will flow from the site according to the program and site conditions.

- **Program**—This screen describes the man-made elements around which water must be made to flow. These generally include paved elements, structures, and landscaping features, all of which must remain undisturbed by the contour modification. You are also told that the regraded site is limited to a specific maximum slope.

- **Tips**—This screen tells a candidate very little. It notes that the *erase* tool affects all changes made to a contour, while the *undo* tool affects only the last action. The most important tip that is inexplicably missing concerns the *move, adjust* tool, which is the one tool that is vitally needed to revise existing contours.

The preprinted site plan, on which you must present your solution, is found on the work screen, which you can access by touching the space bar on the computer keyboard. The site indicates the bordering property lines, complete contours, landscape features, and man-made elements. Nothing more is shown, nor is anything more necessary to solve this problem. The abbreviated tool icons are shown along the left side of the screen, since there is essentially only one tool that will be used to solve the problem.

DESIGN PROCEDURE

The Site Grading vignette requires a candidate to modify the contours on a site in order to drain the surface water in a specific manner. Generally, there will be a paved area and some trees that must remain undisturbed, while the contours are manipulated so that the surface water is directed around them. Therefore, while the title of this vignette is Site Grading, it might just as well have been Site Drainage.

Although none of the vignettes on the Site Planning examination is easy, solving the Site Grading problem requires more technical knowledge than any other vignette. One could not even begin a solution to such a problem without knowing the principles of contour manipulation for the purposes of grading and drainage. Some of the important principles are listed below:

1. To begin with, a *contour* is an imaginary line that connects points of equal elevation. If, for example, a contour were labeled 125, every point along that contour line would be at the identical elevation 125.
2. *Existing contours* are generally shown with a dashed line, while modified or *finish contours* are shown with a solid line.
3. The *contour interval* is the uniform difference in elevation between adjacent contours. In most vignette problems, the contour interval will be one or two feet.
4. Contours that are close together represent a *steep slope*, whereas contours further apart represent a *shallow slope*.
5. Contours that are evenly spaced represent a *uniform slope*. This is an important configuration when creating a consistently sloped driveway, parking area, or pedestrian walk.

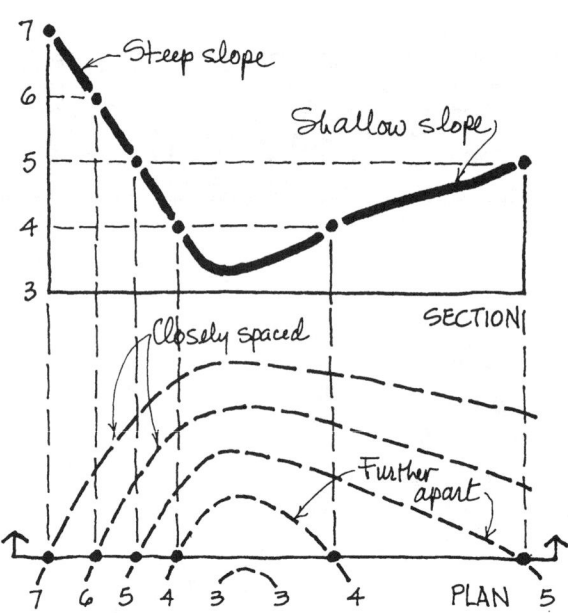

VALLEY CONFIGURATION

6. Contours that point uphill represent a *valley*. This is an essential configuration when solving drainage problems, because one creates valleys, or *swales*, to drain surface water. A common error on this vignette is for a candidate to mistakenly point the contours downhill. This creates ridges instead of swales, sheds water instead of channeling it, and always leads to a failing grade.
7. Surface water always flows in a direction perpendicular to contours.

A *swale* is a small valley-like configuration used to drain surface water in a specific direction. One should try to maintain uniformly spaced contours in a swale to assure that the drained water flows at a consistent speed. Swales are generally depicted as steep, open triangles with a narrow rounded end pointing uphill. Using the ARE computer program, however, the uphill end of each loop will always come to a point.

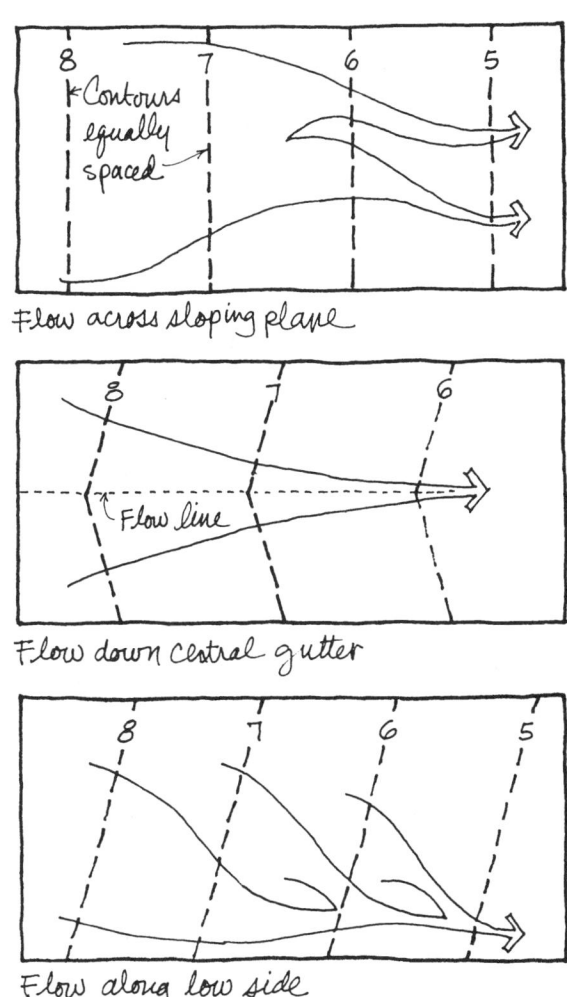

DRAINAGE OF PAVED SURFACES

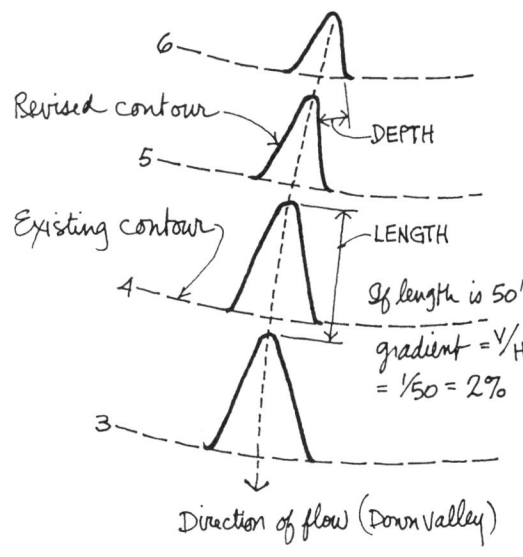

DRAINAGE SWALE

To drain surface water around a level area, a contour of a lower elevation must essentially surround the entire level area. For example, if the elevation of a terrace were 15, then the 14 contour would be drawn behind the terrace, and swales would be created to drain the surface water around the entire terrace. The lower elevation contour behind the terrace actually functions as a drainage ditch.

Since this is essentially a problem of creating drainage swales, one must become proficient in using the *move, adjust* tool to achieve this goal. One proceeds in the following way:

- On the Site Grading site plan screen, click on the *move, adjust* tool. This will bring up a series of small squares that are centered along the length of every contour line. These squares are randomly spaced with apparently no relationship to one another.

- Using the mouse, place the cross-shaped cursor over any of these squares and click once.

- Clicking activates the square, which may then be dragged to a new position on the site. The part of the contour that is repositioned is that portion that lies between the two adjacent squares on the contour line.

- Click again and the modified contour remains set in its new position on the site.

- One may also reposition any square along the contour line itself. This is done in the same way as just described. The purpose of moving a square along its length is to control the shape of a contour in a swale, to make its wide end narrower or more open.

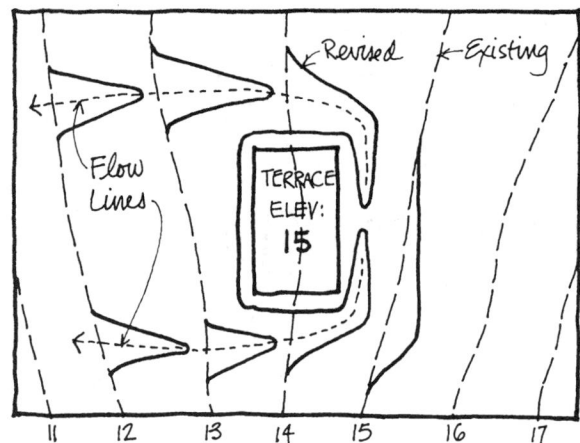

DRAINING SURFACE WATER
AROUND A LEVEL AREA

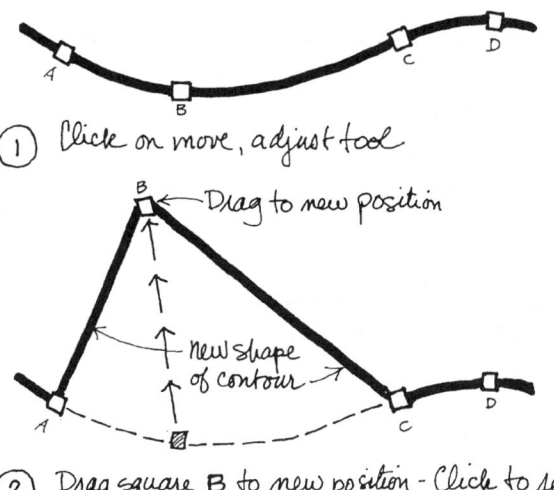

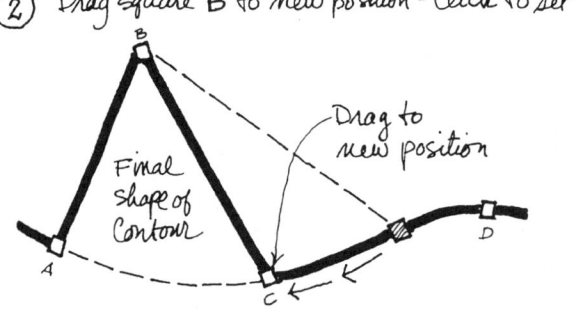

MODIFYING A COMPUTERIZED CONTOUR

With some practice, candidates should be able to modify contours quickly and accurately. Not every contour, of course, will require modification. You should change only those that affect the drainage of surface water around the fixed landscape or man-made elements. Once the surface water has passed the element, the swale can end and the water will continue to drain downhill in an unguided fashion.

An important part of constructing drainage swales is controlling the slope of the regraded portions of the site. First, one must be aware of the maximum allowable slope. For example, the program might state that *regraded slopes may not exceed 20 percent*. In that event, modified contours with a contour interval of one foot would be spaced a *minimum* of five feet apart. The formula one uses is as follows: The Gradient = Vertical Component ÷ Horizontal Component, or in this case 1' ÷ 5' = 0.2 or 20%.

Having determined the minimum allowable spacing between contours, such as five feet in the example above, you must then be certain that your regraded contours do not exceed that dimension. This may be verified with the use of several *sketch* tools. Using the *measure* and *line* tools, distances may be determined directly and observed in the element information area at the bottom of the screen.

Another and perhaps quicker way to verify contour spacing is as follows: Use the *circle* tool to construct a circle with a radius of 2.5 feet (a diameter of 5 feet), which is an amount that can be read in the element information area. That circle may then be used repeatedly to check the distance between contours at any number of points along a contour.

Some candidates may find it helpful to construct a swale flow line before actually constructing the swale. One may use the *sketch line* tool to draw the pattern of flow around the vignette elements and down the site. This flow line then becomes the central spine on which the modified contour loops are arranged.

KEY COMPUTER TOOLS

- **Move Tool** Critical to this vignette is the move tool. The candidate is advised to become very familiar with this tool in the practice exams as it can be a tricky tool to use.

VIGNETTE 5 SITE GRADING

A proposed parking plan has been superimposed over an existing site. You are to regrade the site and modify existing contours so that all surface water will flow to the storm drain inlet shown.

1. Draw revised contours over the paved parking lot so that surface water will be directed towards the storm drain inlet.
   - The 61 finish contour has been established at the paved driveway.
   - A 57 finish spot grade has been established at the north end of the paved parking area.

2. Draw revised contours on all sides of the paved parking area so that surface water from unpaved areas will not run over the paved parking area.
   - Direct all surface water toward the storm drain inlet.

3. Additional requirements:
   - Finish contours shall be continuous from one end of a property line to another.
   - The regraded slope on the paved parking lot shall not exceed 5 percent.
   - The regraded slope outside the parking lot shall not exceed 20 percent.

Lesson Eight: Site Grading Vignette

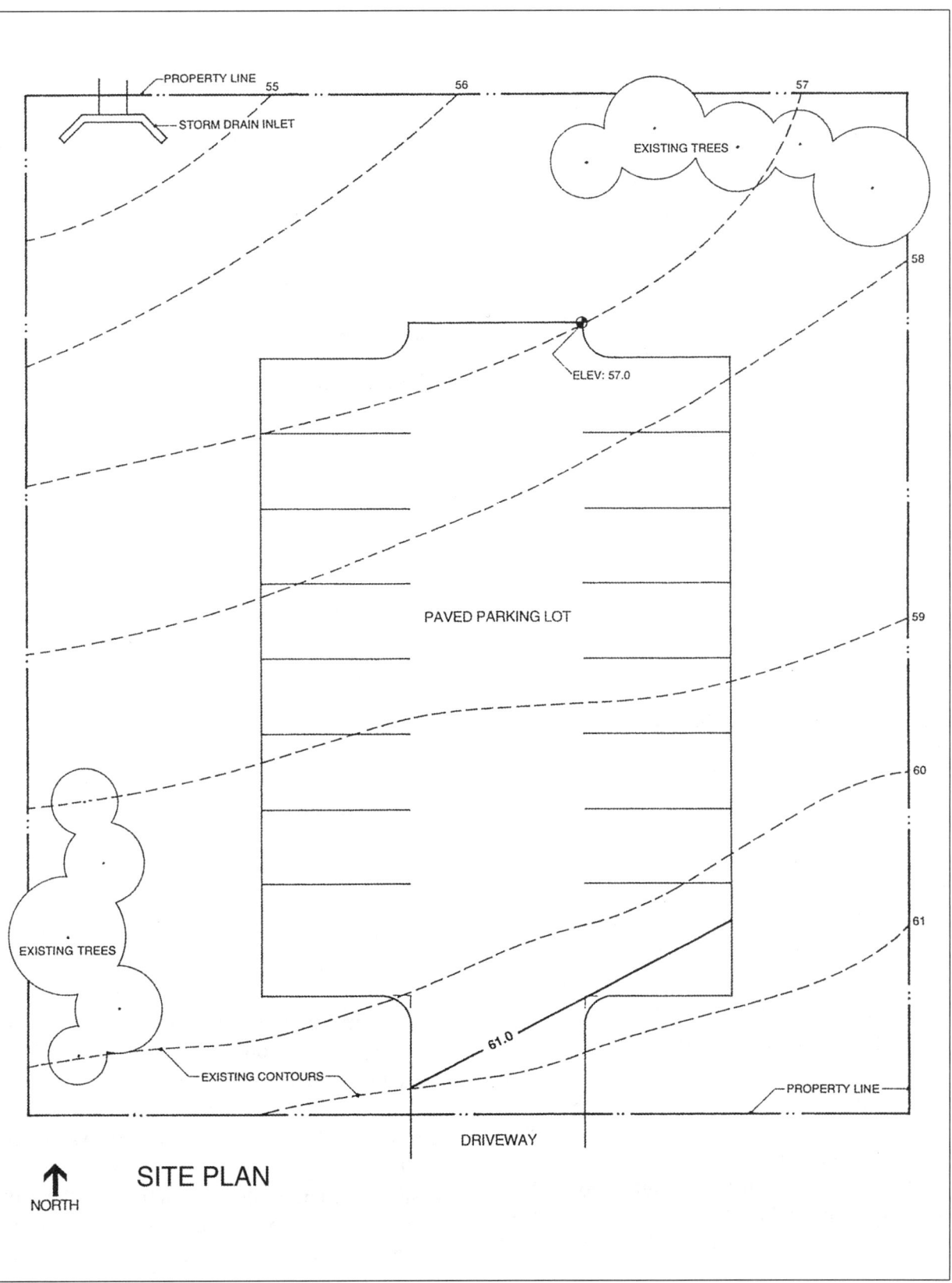

SITE PLAN

VIGNETTE 5   SITE GRADING

Introduction

The following Site Grading vignette was originally part of an ALS paper-and-pencil Site Planning Mock Exam. It has many of the same elements and details found on the actual exam. What is more important, it will help you understand the connection between contour modification and the drainage of surface water from a site.

The Exam Sheet

Shown on the preceding pages are the program and site plan on which candidates were to present their solutions. The scale of the original drawing was 1" = 16', but it has been reduced here to fit the course format.

The program describes a parking arrangement that has been superimposed over an existing site. You are asked to regrade the site and modify the existing contours so that surface water will flow to the storm drain inlet. Specifically, you are instructed to revise the contours over the paved area so that water is directed toward the drain.

You are further instructed to revise the contours around the paved parking area so that surface water from the unpaved areas will not run over the paving. Finally, we are told that regraded slopes may not exceed 5 percent at paved areas and 20 percent at other portions of the site.

The site plan shows a paved parking area for 16 cars superimposed over an existing rectangular site. The grade drops over six feet in elevation, from a little over 61 where the driveway meets the south property line, to the storm drain inlet at the northwest corner of the site. There are two existing tree clumps, at the northeast and southwest, and it must be assumed that these will remain undisturbed. The existing contours have a one-foot interval, and to get you started, you are given the revised contour 61 near the driveway and a spot elevation 57 at the north.

Regrading the Parking Lot

There are several things we know about the parking lot. First, its finish grade may not exceed five percent. Second, the surface water must be drained in the northwest direction, toward the storm drain inlet. Finally, the first revised contour of 61 has been given to us. Therefore, the pattern of revised contours is established by contour 61, and all subsequent contours should be parallel.

With a slope limit of 5 percent (0.05) and a contour interval of one foot, we apply the formula $G = V \div H$, where the gradient G = the vertical component $V \div$ the horizontal component H. Solving for H, we get $H = V \div G$, or $H = 1 \div 0.05 = 20$ feet. Thus, the contours crossing the paved parking area must be 20 feet apart to maintain a uniform 5 percent slope.

Using the same angle established by contour 61, we lay out one-foot contours across the parking area that are exactly 20 feet apart. Because of the vignette design, contour 57 intersects the given spot elevation, and this should send the signal that we are on the right track.

Our next task is to drain the surface water from the land surrounding the paved parking area. We do this by creating swales that begin at the driveway, run along the east and west sides of the parking area, and continue to the storm drain inlet.

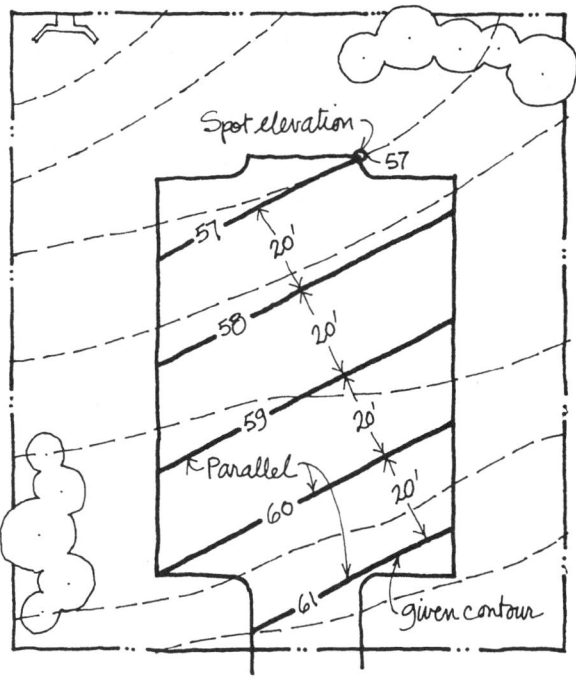

REGRADING THE PARKING LOT

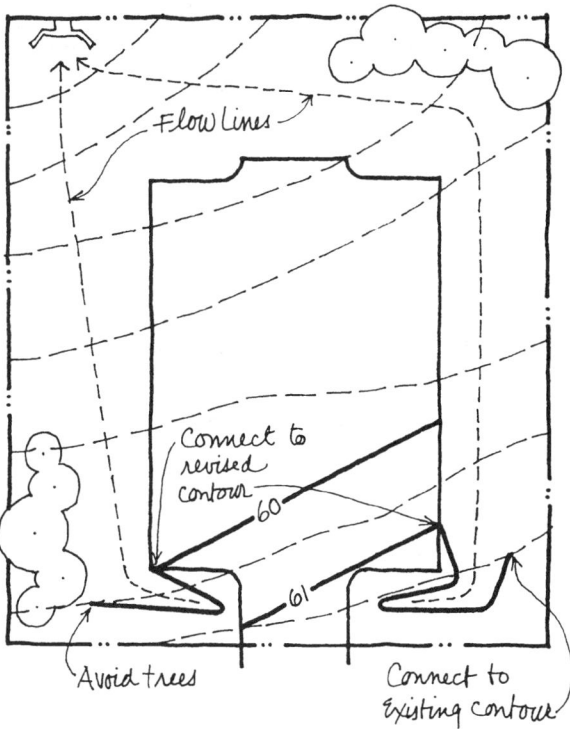

CREATING NEW SWALES

We begin by drawing flow lines that will act as guides and centerlines for the swale loops. Starting at the uphill part of the site, we reconfigure contour 61 with a loop that runs nearly to the paved driveway. One end of the loop turns north and reconnects with the revised 61 contour running across the paved area. The other end of the loop widens a bit and connects with the existing 61 contour.

We have just created a drainage ditch that will catch the surface water from higher elevations, guide it around the southeast corner of the parking area, and prevent it from flowing across the paved parking area. Our configuration is that of a small valley (contour pointing uphill), and with this first modification, we have set the pattern for the remaining modified contours.

The first loop of the swale on the western side is configured in a similar way. One end of the loop connects with the 60 contour at the southwest corner of the paved area, and the other end of the loop connects with the existing 60 contour near the tree clump at the southwest corner of the site.

The remaining contours on both sides of the parking area are modified in much the same way as the 60 and 61 contours. Each modification begins where a contour meets the edge of the paved area, loops around the dashed flow line, and reconnects with the same contour. There are two details to keep in mind while completing this exercise. First, you must be certain which contour you are working on, because it is easy to become confused and lose your way. Second, you should maintain a similar distance between the loop ends, so that surface water will flow at a consistent speed.

Finally, you must verify that no regraded portion of the site exceeds the maximum slope of 20 percent. A 20 percent slope with a contour interval of one foot means that contours should be spaced no closer than five feet apart (G = V ÷ H or 0.20 = 1 ÷ 5). Except for the actual swales themselves, that is, the land within the drainage areas, the appropriate distance has been maintained.

A review of the completed drainage scheme shows the swales conducting water from the highest portion of the site to the storm drain inlet. The rain water that falls on the paved parking area runs in a northwesterly direction toward the same inlet. Therefore, all necessary contours have been modified within the site limits, the surface water has been properly redirected toward the storm drain inlet, and no trees have been affected. Incidentally, the portion of each contour that was not modified is now considered part of the finish contour.

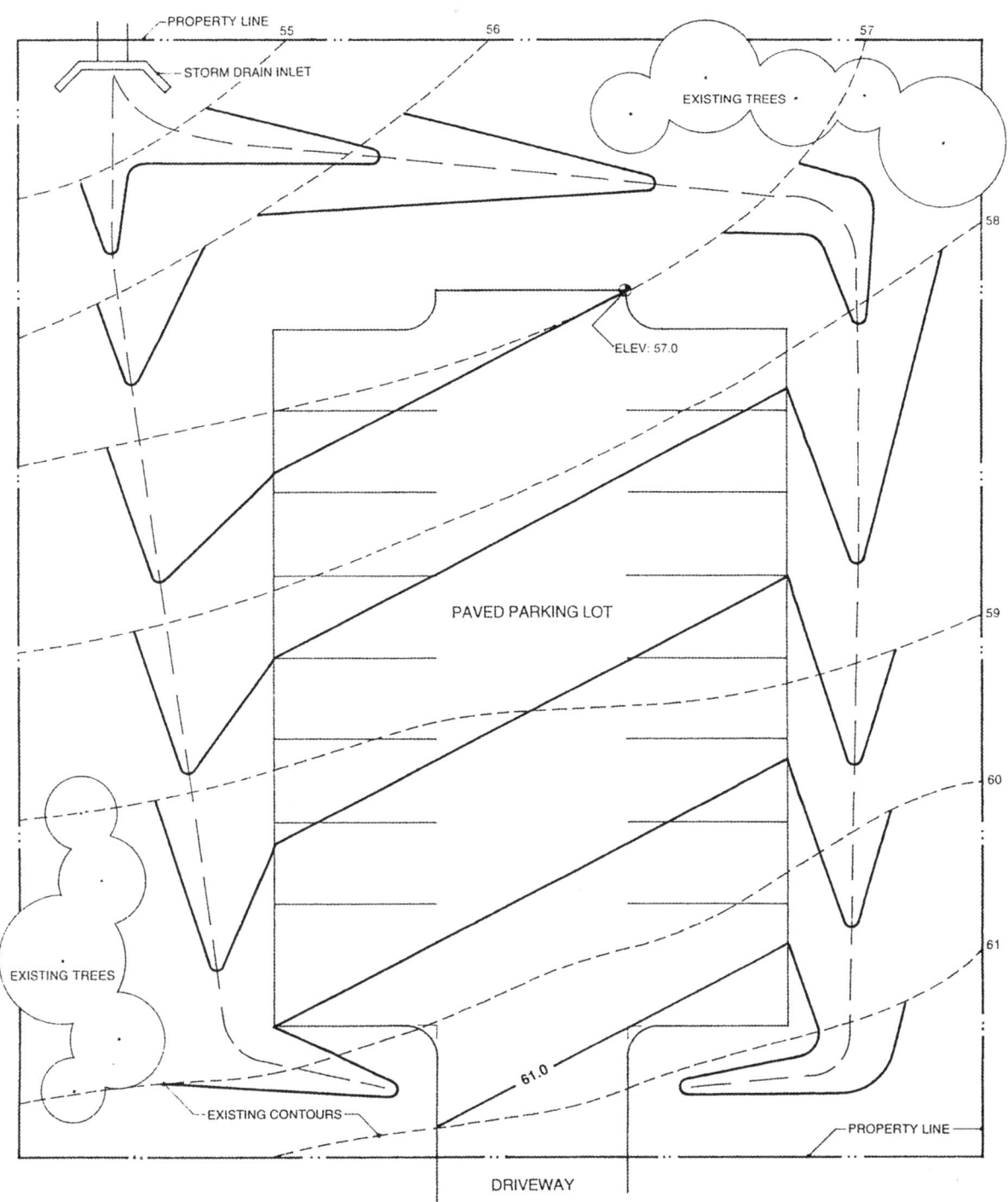

SITE GRADING VIGNETTE - SUGGESTED SOLUTION

LESSON NINE

PRACTICE VIGNETTES

Vignette 1 Buildable Area
Vignette 2 Parking Layout
Vignette 3 Site Parking
Vignette 4 Site Grading

Prior to the current computer exam, Site Planning vignettes were solved the old fashioned way, by using pencil and paper. The problems were relatively simple exercises that tested individual site planning issues, and candidates were given about 20 minutes to solve each vignette.

The following three vignettes, taken from actual past exams, are included here to illustrate the process of arriving at a passing solution. Although they exemplify a former exam, we believe they remain relevant for today's candidates.

The vignettes deal with Buildable Area, Parking Layout, and Site Grading, and each can be solved in the allotted time, if one is familiar with the concepts involved. Candidates may wish to solve these problems before reviewing the suggested solution and discussion that follows each problem.

- **Buildable Area:** These vignettes are similar to what the applicant will encounter in the Site Zoning Vignette, which deals with zoning regulations, buildable area limits, and environmental concerns, as well as calculating a building envelope section.

- **Parking Layouts:** Prior to ARE 3.1, a separate parking lot vignette was part of this division. Now the exercise is included as part of the Site Design Vignette. However, it is still helpful to solve a parking lot only vignette in order to think through parking lot circulation, aisle width, how to avoid dead-end parking, and so on. There are two parking lot layout vignettes included in this lesson for that purpose.

- **Site Grading:** Unchanged for ARE 3.1, this site grading practice vignette should be similar to what you will find on the actual exam.

The approach to any vignette problem involves a sequence of design steps, which is outlined below. Candidates are advised to review each of these carefully, as they represent a fundamental strategy that should be clearly understood. These design steps, incidentally, apply to the current computer vignettes, as well as to the examples that follow.

1. **Read the problem statement.** It may be necessary to read the problem statement again until you understand exactly what the solution demands. Never begin to solve the problem until you are aware of every constraint.

2. **Don't modify the program.** Assume the normal interpretation of all program requirements, and follow them exactly. Never read anything more into a problem than what is actually stated or may reasonably be inferred.

3. **Review the printed drawing.** In all cases, this will be a plan or section on which you will superimpose your solution. As you

review the drawing, refer to the problem statement. Note the scale, north arrow, property lines, setback lines, and all other restrictions that will affect your solution.

4. **Make an initial sketch**. Begin with the specific requirements and lay out a rough solution. This may be done on the actual plan or section, or if more complicated, on a piece of sketch paper. If time is running out, draw directly on the final drawing.

5. **Complete the final drawing**. This is the completed solution on which your grade will be based. Refine all sizes, relationships, adjacencies, etc., and complete all details.

6. **Review the program**. When your solution is completed, refer to the program as a final checklist to be certain that you have complied with every requirement.

7. **Keep going**. When a problem is completed, dismiss it from your mind as quickly as possible and move immediately to the next vignette.

VIGNETTE 1 BUILDABLE AREA

Outline the maximum buildable area on the lot shown to the right. Your solution must meet the following conditions.

> Front yard setback . . . . 30'
> Side yard setback . . . . 10'
> Rear yard setback . . . . 30'
> Wetlands maintenance easement . . . . 30'
> Utility easement @ east P̷L . . . . 40'

Discussion of Solution

It is unlikely that you will see a more elementary problem than this example from a few years ago. That is not to say it is easy; however, those who understand it immediately should probably complete it in under 10 minutes.

Following the sequence of procedures previously outlined, we first review the printed drawing. The scale is noted both graphically and by the figures 1" = 40'. Below the scale is a north arrow pointing towards the top of the page. We see a street at the south and four property lines, which form a rectangle. At the northwest corner is an irregular boundary, indicated by straight, dashed line segments, which denotes an area referred to as "wetlands." Finally, along the east side we see a solid line representing an 8-inch sanitary sewer.

The next step is to read the problem statement. Our assignment is to determine the maximum buildable area of the property, and this is accomplished by graphically interpreting each of the five distinct restrictions and applying those that are most limiting. This problem is simple enough so that our first sketch might very well be drawn directly on the final sheet.

Where several conditions apply, it is always advisable to deal with the restrictions consecutively, in the same order as they appear in the problem statement. This will help avoid a careless oversight or omission. For example, the first restriction is the front yard setback, which is noted as 30 feet, and therefore, we begin with this setback. In this problem, the front of the site is obvious; it is the southerly side of the property facing the street. However, we see three lines here: the street centerline, the curb line, and the southern property line. Candidates must realize that all yard setbacks are measured from property lines. Therefore, we draw a straight line parallel to the property line and 30 feet from it. This new line is a graphic representation of our first restriction.

The next restriction is the side yard setback of 10 feet. These limit lines are drawn on both sides, parallel to the east and west property lines. In a similar manner, the rear yard setback of 30 feet is drawn parallel to the northerly property line. So far this vignette has been fairly easy. In fact, it has been so simple and straightforward that candidates might wonder about all the horror stories they have heard concerning the difficulty of this test. The following two restrictions may help answer that question.

We next interpret the restriction noted as a 30-foot wetlands maintenance easement. First of all, one might ask, what are wetlands? They are land areas that are low, poorly drained, and often have standing water. They are also known as marshes, swamps, or bogs, and they are invariably unsuitable for construction. If you assumed that this was simply a wet area that was so undesirable that you had to stay 30 feet away from it, you were absolutely correct; and in fact, you need not know any more than that. The graphic interpretation of the 30-foot wetlands maintenance easement, therefore, is a line 30 feet from, and parallel to, each segment of the wetlands boundary line.

The final restriction is the 40-foot utility easement at the east property line, which is a 40-foot-wide strip of property reserved for the placement and maintenance of utility lines. But where should it be located on our site? There are only three reasonable possibilities: (1) centered on the east property line, (2) centered on the 8-inch sewer line, or (3) placed at the most easterly 40 feet of the property.

The clue here is found in the location of the sewer line. That line, which represents the centerline of the sewer, scales 21 feet from the eastern property line. Therefore, if the utility easement were centered on the property line, the sewer line would not even fall within the easement. Similarly, if the easement were centered on the sewer line, the east side of the easement would not even reach the property line, where it should be, according to the program. The only choice remaining is to place the utility easement so that its easterly edge coincides with the easterly property line and its westerly edge is parallel to and 40 feet from the property line.

With all of the required setback lines drawn, we must now connect them in order to determine the most restrictive shape within which one can legally build. As you can see, the easterly side yard setback is irrelevant, because it is superseded by the utility easement, which is more restrictive. Similarly, portions of the westerly side yard setback and rear yard setback are superseded by the wetlands easement, which again is more limiting.

A final detail is the small piece of limit line situated opposite the most southerly segment of the wetlands boundary. This slight nick off the corner is caused by the wetlands boundary 30 feet to the west, and it is actually a minor fragment of the overall outline that could easily be overlooked. Ignoring this detail would probably not result in a failing grade, but this is exactly the kind of unforeseen condition that candidates must watch out for.

Finally, we outline the maximum buildable area with a heavy line, so there will be no confusion among the construction lines that may remain. Although it was not specifically required, we have indicated all key setback dimensions to make it easier for graders to verify the accuracy of our solution; and with that, we move quickly to the next vignette.

VIGNETTE 2 PARKING LAYOUT

Given

A neighborhood church requires overflow parking on a lot across the street. A 25 ft [7.9 m] curb cut from Mission Street is shown. Sixty-five spaces are needed that are each to be 9 ft [2.9 m] wide and 20 ft [6.3 m] deep with a minimum 25 ft [7.9 m] aisle. No dead-end parking is permitted. (Handicapped parking is provided elsewhere.)

As shown below, every individual row of spaces must be bordered with a minimum 5 ft [1.6 m] wide planter strip and no more than 14 spaces may occur contiguously without an intervening 5 ft [1.6 m] wide planter strip.

Setbacks are shown on the site plan. No parking space or driveway shall be built in the setback areas. However, planter strips and access drives may occur in the setbacks.

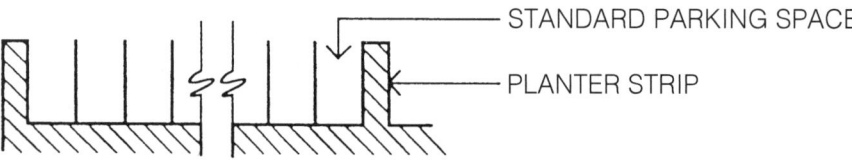

NOT TO SCALE

Assignment

Draw the parking layout for the church lot, showing all the spaces and planter strips. Number each space.

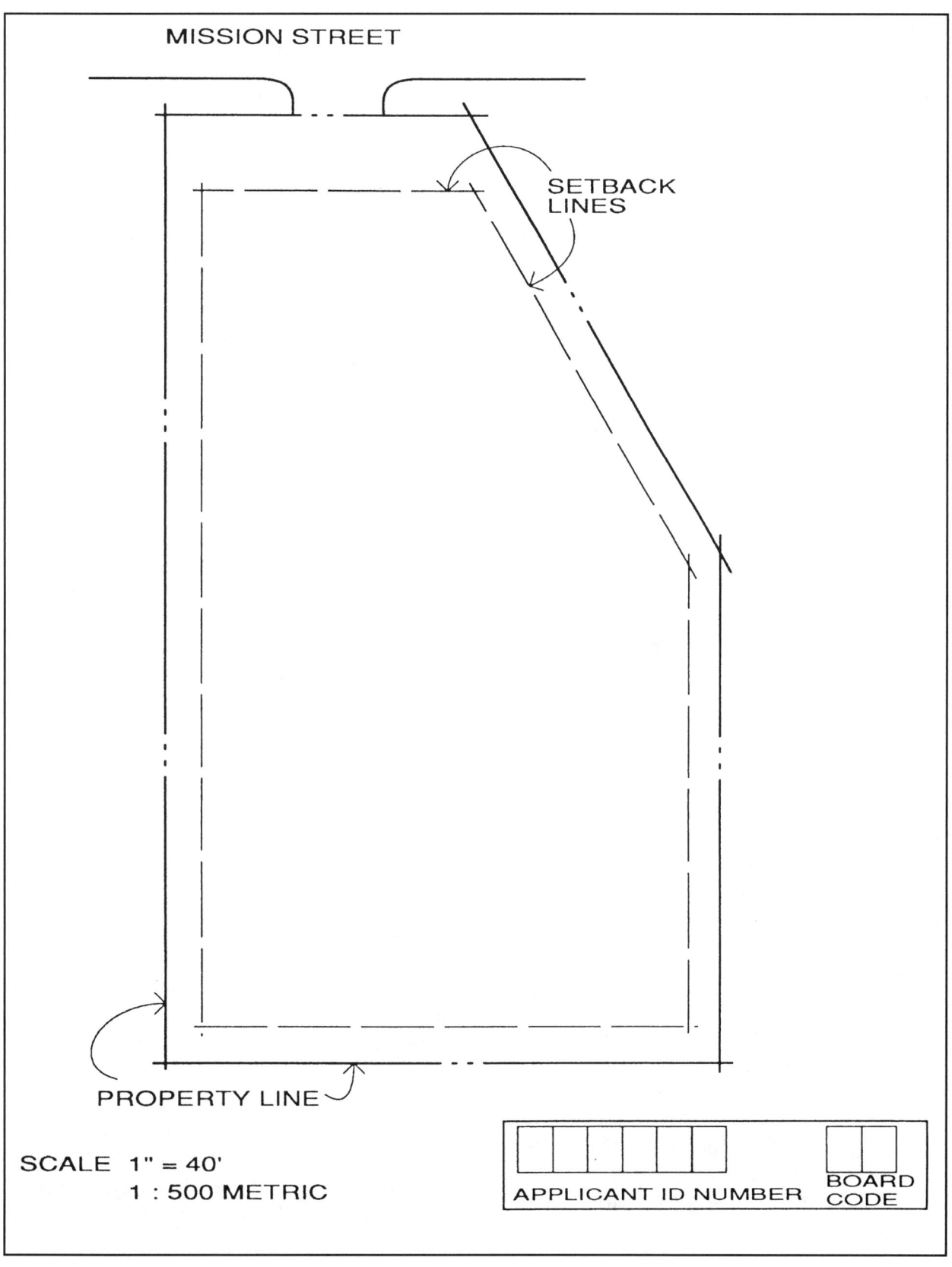

Discussion of Solution

This is another relatively simple problem, provided of course that you are familiar with the fundamentals of parking layouts. Once again we begin by reviewing the printed drawing. No north arrow has been indicated, but for purposes of this discussion, we assume north to be at the top of the sheet. We see a rectangular site with the northeast corner sliced off at a 30-degree angle. Mission Street is shown at the north, and setback lines are indicated at all five sides of the geometric figure. Finally, the drawing scale is noted at the engineering scale of 1" = 40'.

The next step is to read the problem statement, which describes the site as a location for overflow parking for a neighborhood church located across Mission Street. The 25-foot-wide curb cut is described, as is the need for 65 car spaces that are each 9 feet by 20 feet in size. We are also told that circulation aisles are a minimum of 25 feet wide, and dead-end parking is prohibited.

For those unfamiliar with the parking term *dead-end*, it refers to a circulation layout in which cars are unable to circulate continuously from the entrance to the exit of the parking area. In other words, at some point, a driver is forced to put the car in reverse and turn around to exit the site. In this, and most other vignette problems, cars must be able to enter and exit the site while always moving ahead in a forward gear.

The program also includes the requirement that individual rows of spaces must be bordered with a minimum 5-foot-wide planter strip, and no more than 14 spaces may occur contiguously without an intervening 5-foot-wide planter strip. The accompanying illustration shows a few typical parking spaces with the desired planter strips. Candidates should note that the planter strip borders three sides of the typical parking row.

Finally, the program states that no parking spaces or driveway may encroach into the setback areas, but planter strips and access drive may occur in the setbacks.

Before beginning to draw, we should state a few general observations. This parking lot serves a church across Mission Street. Therefore, to be as close as possible to the facility being served, cars should be parked close to Mission Street. In addition, the most spacious part of the lot is at the west side, away from the angled northeast corner. Most cars, therefore, should be parked along the west setback line. To avoid the dreaded dead-end condition, we must create a loop driveway that enters and exits the lot at the indicated Mission Street curb cut.

These observations are the guidelines by which we will determine a quick and effective solution. Candidates who disregard this analysis and rush to lay out cars in a hit-or-miss fashion may very well run out of time before finding a solution. That is why a few moments of analysis ultimately save a great deal of drawing time.

Our first line of parking begins at the intersection of the north and west setback lines. Car number one is placed tight against that corner, because the adjacent planter strips are permitted to be outside the setback lines. We lay out the maximum permitted 14 spaces towards the south and then create a 5-foot-wide planter strip. We continue laying out cars along the same row, but because we are getting further away from the church, the exact number will have to wait until we see how many cars fit elsewhere.

The remainder of the cars will be fitted in parallel rows east of this first row. From west to east, the rows will lay out as follows: The original parking row, a 25-foot-wide driveway, the second parking row, a 5-foot-wide planter strip, the

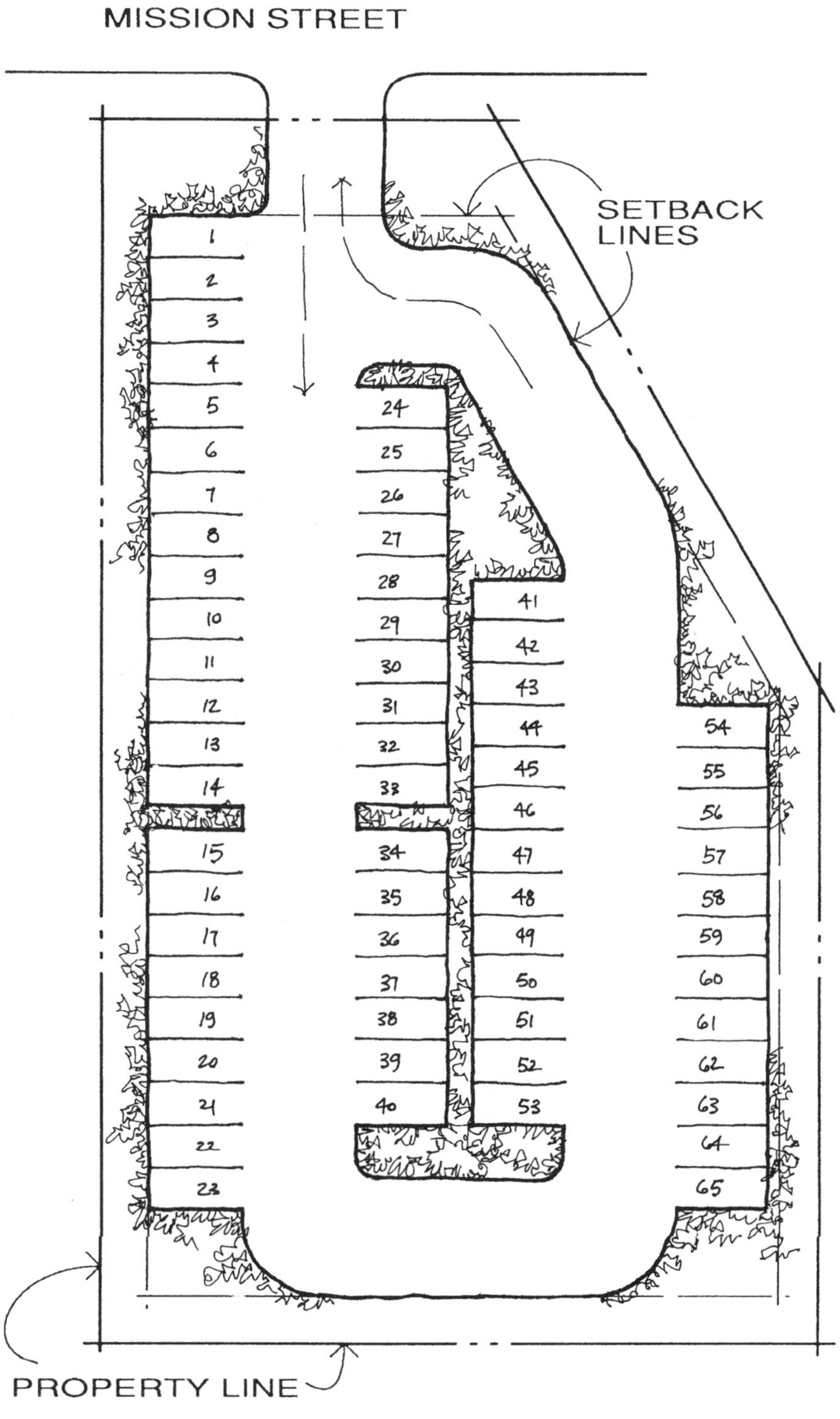

Lesson Nine: Practice Vignettes

MISSION STREET

SETBACK LINES

PROPERTY LINE

SCALE 1" = 40'
1 : 500 METRIC

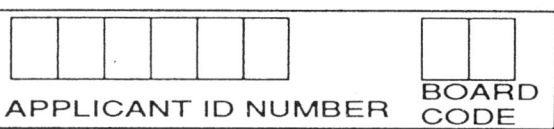
APPLICANT ID NUMBER BOARD CODE

third parking row, another 25-foot-wide return driveway, and the fourth and final parking row. The total width of the four parking rows, two driveways, and one planter strip is $(4 \times 20) + (2 \times 25) + 5 = 135$ feet. Since the distance between the east and west setback lines scales about 138 feet, this arrangement will just fit.

We next create the driveway loop that runs south from the Mission Street curb cut to the south setback line. At that point, it loops back northward until it hits the angled setback line, runs along that line, and then returns to the Mission Street curb cut.

Car number 24 begins the second parking row and is located five feet south of the northern driveway section to allow for the planter strip. We arbitrarily place the intermediate planter strip, as well as the cars in this row, opposite the ones in the first parking row. The row ends with car number 40, which is separated from the south driveway by a planter strip that is about 11 feet wide.

The third parking row contains 13 cars, and so an intermediate planting strip is not necessary. The fourth parking row begins with the last car, number 65, and matches the car layout of the first row. It runs north to car number 54, which is the last car that fits along the east side.

A certain amount of trial and error went into the final layout. There might have been one more car in this row or that, but all passing solutions would resemble the one shown. It was preferable to avoid parking perpendicular to the angled setback line, because that additional complication would have consumed too much time.

To complete our solution, the spaces were numbered, as requested by the program. Although the planter strips in the program were shown hatched, we have added a texture that was applied much more quickly.

VIGNETTE 3 SITE PARKING

Introduction

The following example of a Site Parking problem came from a recent ALS paper-and-pencil Site Planning Mock Exam. It contains many of the same features one may expect to see on the actual test. As with the other vignette examples, our solution is presented in a logical sequence of steps. Those who learn to solve vignette problems in this way should have few difficulties.

The Exam Sheet

Shown on the following pages are the program and site plan for the Site Parking vignette, which was part of the Mock Exam package. Candidates were instructed to present their solutions directly on the site plan sheet. The original scale of the drawing was 1" = 40', but it has been reduced here to 1" = 50' to fit the course format.

The program requires one to provide a paved parking layout for a small branch Auto Club Office. The requirements are relatively simple and divided into three sections: the required parking spaces, the parking lot circulation, and additional restrictions.

The partial site plan shows a rectangular plot of ground that is bordered on the west by Tow Truck Lane, which runs in a northeasterly direction at about a 10-degree angle to the opposite lot line. The Auto Club Office is seen at the southwest corner of the site, and at the northeast corner we can see a large rock outcrop. The rock outcrop, together with the existing tree clumps, represent elements that must be avoided in our new development.

Locating the Parking Area

Before we can locate the parking area, we must draw every setback restriction that will influence this decision. Among the *additional requirements* in the program, we see that paving is restricted to an area that is no closer to the property lines or to the Auto Club Office than 50 feet. In addition, no paving is permitted any closer to Tow Truck Lane than 70 feet. If we draw these restrictive lines, we can see that a relatively small development area remains near the center of the site.

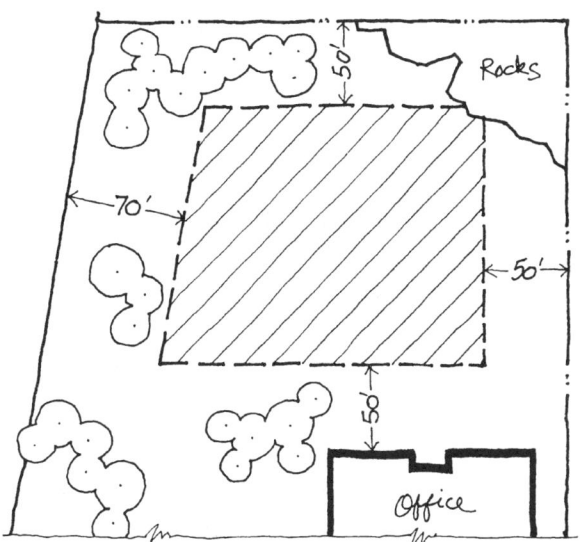

AREA OF DEVELOPMENT

Laying Out the Cars

In any such parking problem, it must be assumed that it is best to locate the cars as close as possible to the building being served. Thus, we lay out the first line of cars along the southern boundary of our development area, because these cars will be closest to the Auto Club entrance. We place the four required accessible spaces at the southeast corner of the development area, because that

VIGNETTE 3 SITE PARKING

On the site plan shown, you are to provide a paved parking layout to serve a small branch Auto Club Office.

1. Design a paved parking layout to accommodate a total of 49 cars as follows:
   - 30 standard spaces (10' × 20') for Auto Club patrons.
   - 4 accessible spaces (15' × 20') for Auto Club patrons.
   - 15 standard spaces (10' × 20') for Auto Club employees. These spaces must be clearly separated from patron parking.
   - All parking spaces must be perpendicular to traffic aisles.
   - Provide 5-foot-wide landscape buffers at the ends of parking rows where they meet driveways.

2. Draw all driveways and traffic aisles required to connect parking to the street.
   - All drives and traffic aisles shall be 25 feet wide.
   - Drive-through circulation is required with NO dead-end parking. Show traffic pattern.
   - Only one curb cut from the street onto the site is permitted to cross the setback area.
   - The intersection of the access drive and Tow Truck Lane shall be perpendicular for at least the first 50 feet of the drive.

3. Additional requirements:
   - Paving shall be no closer to the property lines or the Auto Club Office than 50 feet.
   - Paving shall be no closer to Tow Truck Lane than 70 feet.
   - Show all paving limit lines. Parking is not allowed outside paving limit lines.
   - Paving shall be minimized and shall not disturb existing site trees or other features.

Lesson Nine: Practice Vignettes

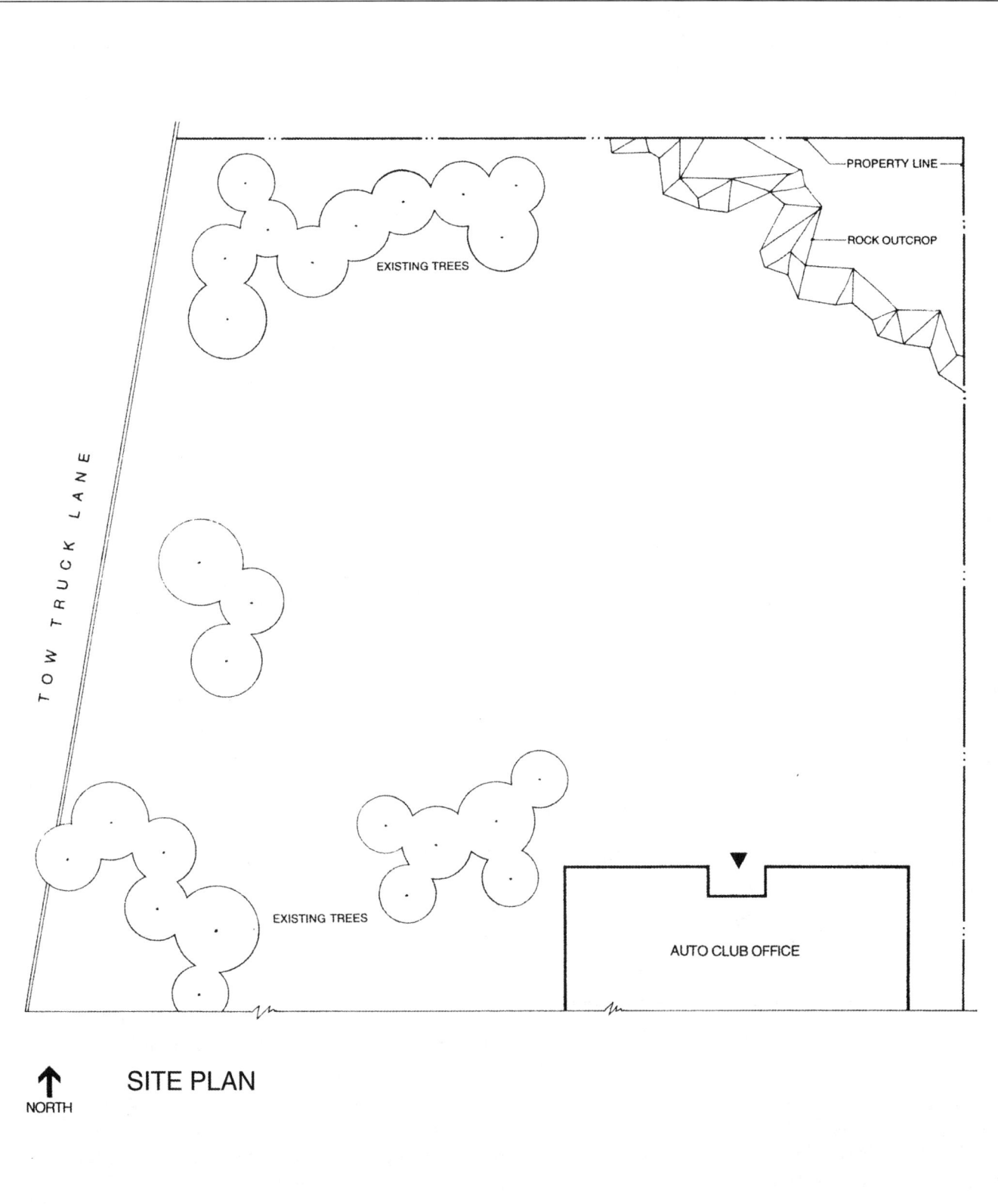

SITE PLAN
NORTH

location is directly opposite and as near the front door as permitted. Moving west along that southern side, we can fit about a dozen more standard spaces.

For efficiency, we have elected to design this parking layout using double-loaded traffic aisles. Thus, we draw a 25-foot-wide traffic aisle adjacent and parallel to the first line of cars, and we lay out an additional line of cars on the other side of the aisle. It becomes obvious at this point that we will need more than one double-loaded traffic aisle. If we allow enough space for maneuvering, we can park only a little more than half of the 49 cars that must be accommodated. It is clear, therefore, that we will require two double-loaded traffic aisles. We know that the entrance to the parking area comes from Tow Truck Lane at the west. Therefore, our plan is as follows: We will circulate from west to east, drive down one traffic aisle, loop around, circulate down the second traffic aisle from east to west, and head towards the exit.

The program states that the 15 car spaces for employees should be clearly separated from patron parking. We assume that these spaces can be more distant from the building entrance than the patron parking. The reasoning is as follows: Patrons drop in for relatively short visits, and they need quick and convenient parking. Employees, on the other hand, park for the entire work day, and a slightly longer walk to the building is only a minor inconvenience. Therefore, we decide to locate the 15 employee spaces at the northernmost section of the parking area.

Vehicular Circulation

At this point, we have made most of the decisions that will enable us to lay out the parking lot. We have two adjacent and parallel double-loaded corridors and have determined that our circulation pattern will be in the form of a large

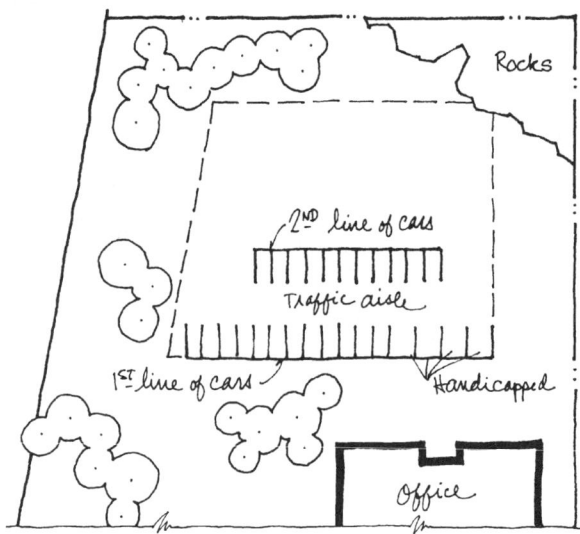

INITIAL CAR LAYOUT

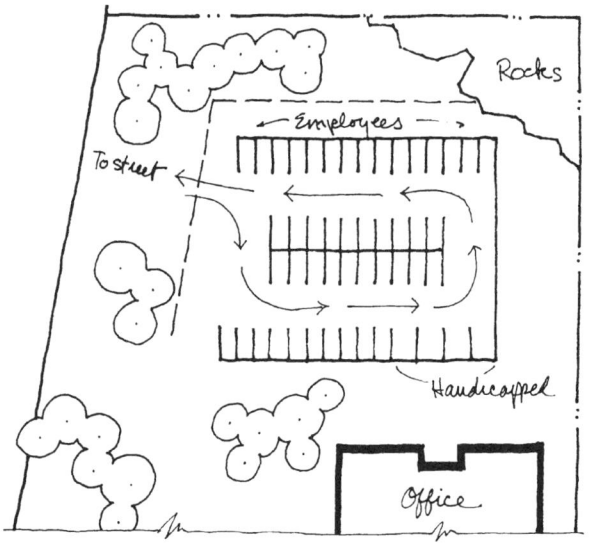

CREATING A LOOP DRIVEWAY

continuous loop—up one aisle and down the other. Earlier we decided that we would circulate from west to east, loop around, and return in the opposite direction to the exit. We believe this is a sound decision; however, without a passenger drop-off to influence our judgment, it is not a pass-or-fail kind of decision.

There is no doubt, however, that traffic would be somewhat more logical and efficient if cars first ran east along the southernmost aisle, which has the most patron car spaces, and is also closest to the building. If no spaces are available here, patrons will loop around to the northern aisle and find a space there. Because the traffic pattern is one-way, employees are forced to circulate through both aisles to reach their parking spaces, but at the end of the day, they will be closer to the exit.

We must next decide where to locate the access driveway from Tow Truck Lane. The program allows only one curb cut, and the driveway must serve both incoming and outgoing cars. The only restrictions here are to make the driveway perpendicular to the street for at least 50 feet and avoid the existing trees. Avoiding the trees is easy, since there is a large gap between the tree clump at the northwest corner and the next clump of three trees south of those.

Although the driveway had to be perpendicular to Tow Truck Lane for at least the first 50 feet, we have made the entire length of the driveway perpendicular to the street. A driveway without bends or curves is easier and safer to maneuver, and we believe the simplified form looks better. However, we should point out that the computer program on the actual exam cares little about aesthetics. A solution that conforms to the programmatic data will generally pass. Therefore, if you drew a driveway with a more complex form, your solution would not be penalized.

Final Adjustments

The final parking lot layout will probably require some tweaking here and there. For example, the program requires 5-foot-wide landscape buffers at the end of parking rows where they meet driveways. These must be joined to the two middle rows of parking before adding the two 25-foot-wide side aisles.

Our final layout shows the 15 employee cars at the far north. The two middle parking rows contain 10 standard car spaces each, and the parking row closest to the building contains 10 more standard car spaces together with the four accessible car spaces. The overall shape of the paving is simple and about as efficient as possible. We have also shown the traffic pattern with directional arrows, so that one can follow a typical car's movement. You can see that the traffic is essentially one-way, and circulation conflicts have been virtually eliminated.

Candidates may wonder about the hazards of pedestrian circulation through a parking area. Unfortunately, this is a fact of modern life. Cars and people coexist in parking lots every day and in every place, and we have all learned how to deal with those situations. As long as the disabled can circulate from their accessible spaces to the building without crossing any roads, you should not worry about the common, and often unavoidable, conflict of cars and people.

In many ways, this was a relatively simple parking layout problem. Once you established the paving limit lines, you were left with a small but adequate area in which to organize the parking layout. It should also be pointed out that the paving limits did not contain any trees and only a small corner of the rock outcrop. Therefore, it was easy to avoid trouble on that score. There were only two breaks in the trees along Tow Truck Lane, and it should have been clear that the northernmost break was the preferred point of access. Finally, one should be ready to apply the lessons learned in this problem, because all such Site Parking vignettes will be similar.

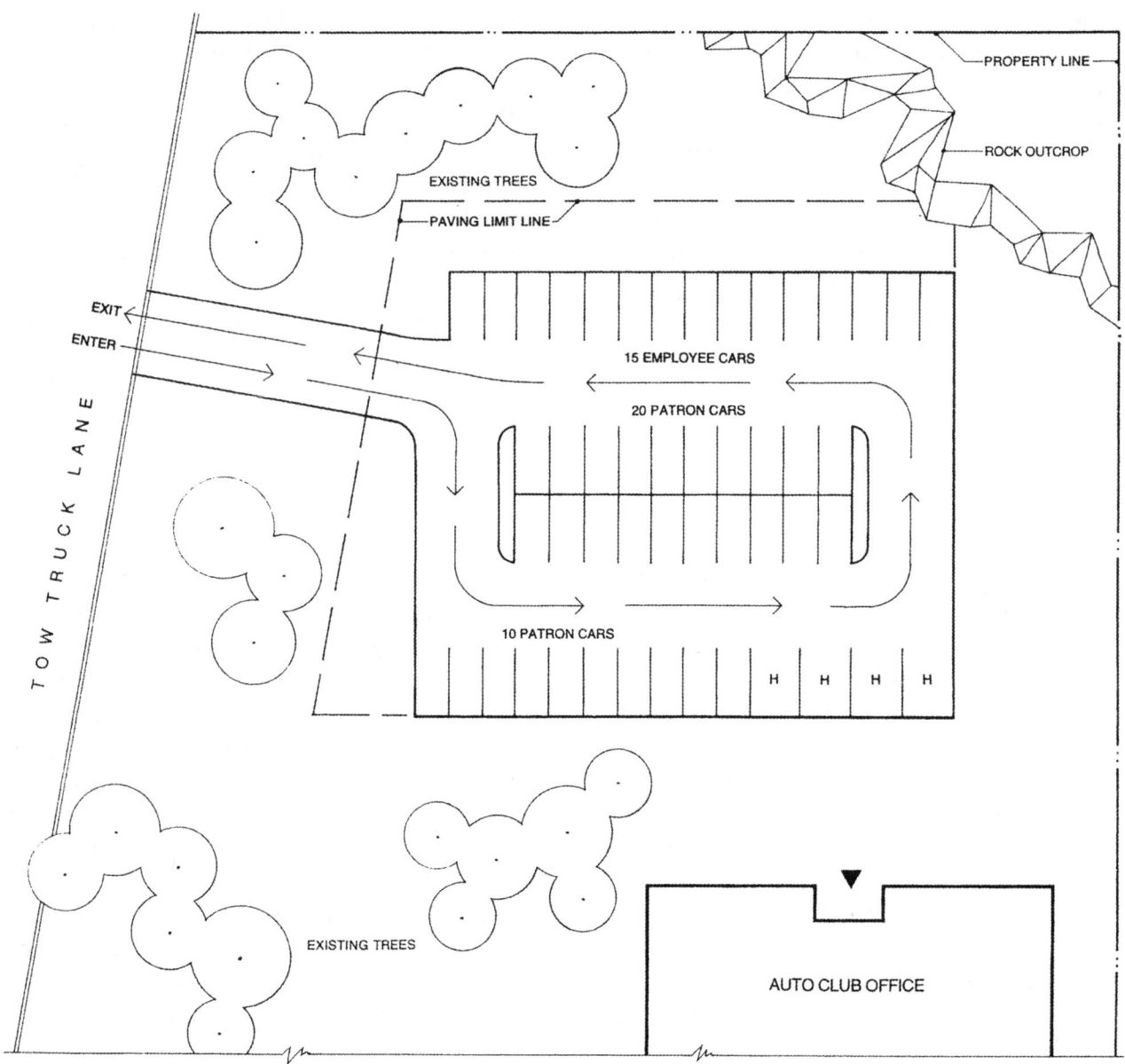

SITE PARKING VIGNETTE - SUGGESTED SOLUTION

VIGNETTE 4 SITE GRADING

Given

- A one story, slab-on-grade residence placed on a lot as shown with the finished floor elevations (F.F.E.) as indicated. A paved driveway is provided to the garage.
- Contours are shown at 2'-0" [0.6 M] intervals.
- Existing vegetation indicated by tree line.
- Underground utilities easement at west property line not to be disturbed.

Assignment

Draw new contour lines required to provide proper access and drainage for total property, including driveway.

- Grade on drive not to exceed 10%.
- Preserve natural vegetation.
- Indicate revised contours as illustrated below.

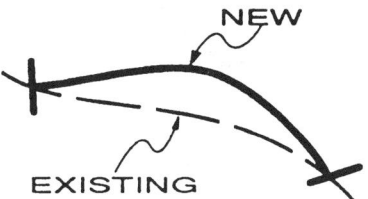

144 Site Planning

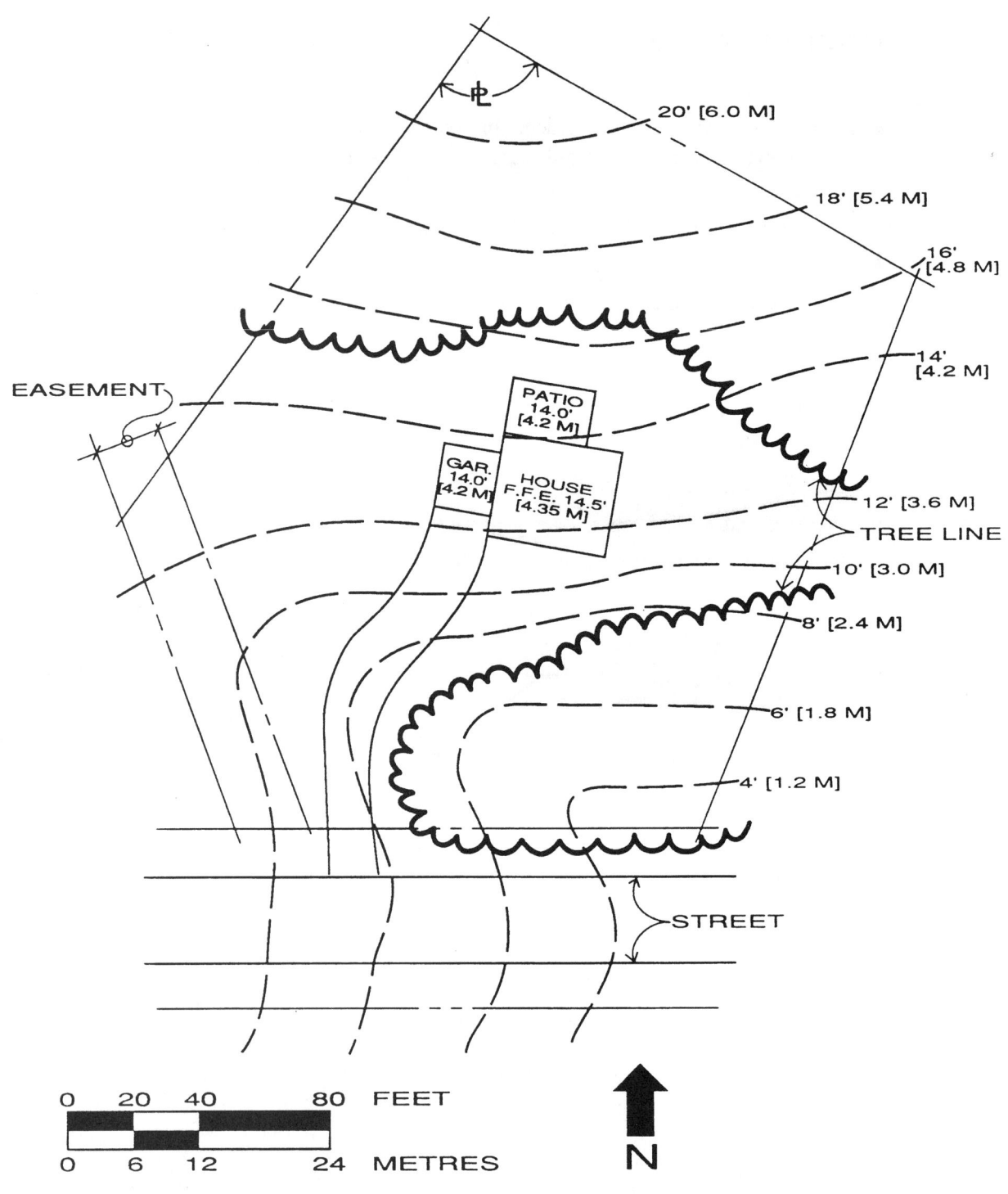

Discussion of Solution

This problem requires more thought, more skill, and probably more time to complete than the two previous problems. In other words, a candidate should have a full 20 minutes of action with this vignette. We see on the printed drawing an irregular lot with a house, garage, and patio nestled among the trees. A curved driveway leads from the garage to a street at the south. There is an easement at the southwest, and contours at two-foot intervals cover the entire site. The north arrow points straight up.

Our assignment is to revise the existing contours to provide proper drainage for the property. The restrictions are relatively simple: the driveway grade may not exceed 10 percent, and both the easement and natural vegetation must remain undisturbed.

A review of the contours reveals that surface water will flow naturally from north to south, and most water will run toward the property's southeast corner. Since we were not advised otherwise, it is apparently acceptable to permit water to flow to the street, where presumably, it will become someone else's concern. Therefore, our only real problem is to guide the surface water around the structures and toward the street.

The devices commonly used for this purpose are drainage swales, which are actually small, graded flow paths that resemble valleys. Like valleys, the topographic configuration of a swale always has the small loop pointing uphill. These looped configurations are created for each contour line, and their centerlines form a smooth channel through which water flows from higher to lower elevations. Before drawing a swale, however, one must first draw a flow line, which begins at the highest point of contour modification and runs in the desired direction of flow. The flow line is actually a spine on which the modified contours are draped.

The revised contours for this problem comprise two distinct parts. First, one must solve the driveway grade, and second, one must create drainage swales around the structures. The modified contours for both parts must ultimately connect and be continuous, because contour lines are always continuous. It is generally best to begin with the architectural elements, since their forms will not change. Thus, we consider the contours crossing the driveway first.

We begin with contour 14, since that is the garage elevation, and we draw a perpendicular line across the driveway about 10 feet south of the garage. This allows a level area just in front of the garage, and the continuation of that line eastward will fall just south of the house, whose elevation is 14.5. Our next line, contour 12, is drawn across the driveway, 20 feet south of modified contour 14, in order to maintain the maximum slope of 10 percent. We determine this distance by using the formula $G = V/H$, where G is the gradient, and V and H are the vertical and horizontal components. Thus, $10\% = 2$ feet $\div H$, or $H = 2 \div .10 = 20$ feet.

Contour 10 is handled in the same way, again with a perpendicular across the driveway that is 20 feet south of contour 12. The next line, contour 8, is a problem, since it does not actually cross the driveway. If we modified contour 8 so that it crossed the driveway, we would be left without a reconnecting line on the west side of the driveway, as shown on the next page.

Perhaps contour 8 could follow the west side of the driveway and then recross the drive to reconnect with the original contour 8, but we believe it is preferable to eliminate the entire problem.

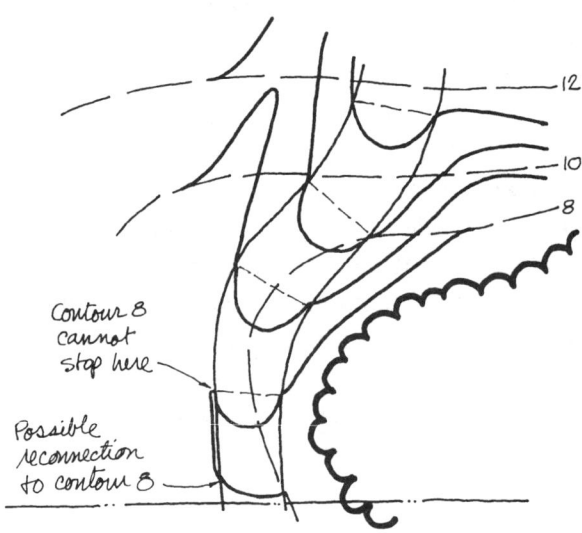

We do so by modifying contour 8 in a way that completely avoids the driveway, except where it reconnects at existing contour 8, near the southern property line.

The problem statement says nothing about the cross section of the driveway, so straight, perpendicular contours are completely acceptable. These represent a straight cross section on which surface water would flow as a sheet down to the street. We have chosen to indicate a crown in the driveway, shown by the contours pointing downhill, which will cause the surface water to flow towards both edges of the driveway. If we had drawn curved contours pointing uphill, it would represent a saucer-like configuration in which water would flow down the center of the drive, which would also be acceptable. To repeat, however, straight contours would have been proper and just as correct as any other configuration.

We must now deal with the swales, and the most important aim of this exercise is to eliminate the possibility of water flowing across the patio or against the structure. This is accomplished by creating a continuous drainage ditch across the rear of the development. Since the patio elevation is 14, we pull contour 14 behind the patio on both sides and create two loops opposite the midpoint of the patio's north side. Beginning at the two loops, we next draw an eastern and western flow line, each of which indicates the preferred pattern of surface water flow. This is the spine on which subsequent modified contours will be hung.

It should be understood that the swale indicated by contour 14 actually shows the two outside edges of a small valley. Within the swale, the elevation is somewhat lower, thereby allowing water to run down from the edges to the central flow line and then continue downhill. Thus, water running down the hill north of the house will enter the swales and follow the flow lines, eventually to the street.

The lower edges of modified contour 14 wrap around the house, across the driveway, and around the garage, creating a form which resembles a moat around an island. The upper edges of modified contour 14, on the other hand, reconnect to existing contour 14 at both sides of the patio.

Similarly, modified contours 12 and 10 follow the pattern established by contour 14. The loop of each begins on the flow line, each reconnects to its existing contour, and all modified contour lines are continuous. It is advisable to keep the horizontal distance between swale loops equidistant, which indicates a continuous and uniform slope. However, the grading program is fairly permissive in this area, so if your swale pattern is not absolutely regular, you would probably not be heavily penalized.

Candidates should note that our revised contours have been drawn about six feet apart at their closest points. Actually, there was no finish grade prescribed for the ground areas; however, we feel that a 3:1 slope (2 foot contours, 6 feet apart) is generally acceptable, and it was easy enough to arrange here. Most hard-pressed candidates rarely have time to give this much thought, and grading is generally lenient in this area. Nevertheless, candidates should understand that contour lines drawn very close together represent slopes which may be exceptionally steep.

With all the relevant contours revised, our solution is complete. However, one may ask, what about the contours we ignored, for example, contour 16 or contour 6? After all, the assignment stated that we should provide proper drainage for the total property. In reality, our problem was to provide positive drainage only as it affected the built elements. Water that falls on the slopes above and below the development will flow toward the street without threat to the structures. Besides, we were told to preserve the natural vegetation, so in no event would one regrade contours beyond the tree lines. Therefore, if your solution resembles what is shown here, you have adequately satisfied this problem.

148 Site Planning

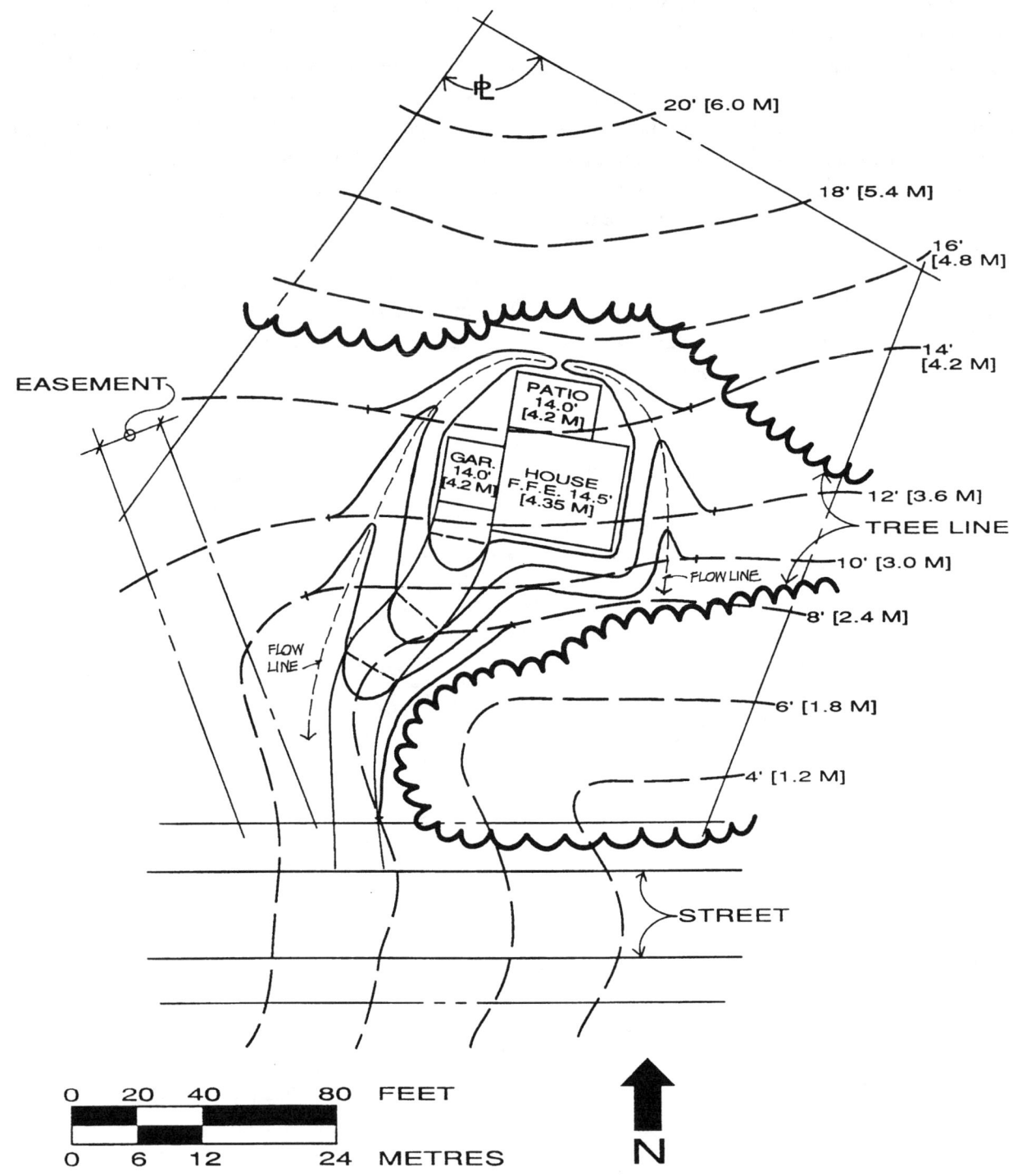

LESSON TEN

EXERCISE PROBLEMS

The following exercise problems, which constitute a major part of this study guide, are intended to expose candidates to the widest possible range of likely site planning subjects. Most of these problems are similar to past exam vignettes, while others involve subject matter that could reasonably appear on future tests. These vignettes are not intended to replicate any particular version of the ARE. However, because each vignette necessitates a quick and practical solution, as a group they provide an effective method for candidates to test their competence in solving a variety of site vignette problems.

Following each exercise problem is a suggested graphic solution together with a complete analysis and explanation. Be aware that our solution represents one way to satisfy the problem presented, not necessarily the only possible way. Whether or not you agree with our analysis or graphic solutions, we feel that they illustrate direct responses to the problems.

In all our commentary, we assume that candidates are reasonably familiar with the fundamentals of site planning. Therefore, some basic theory may be disregarded, on the assumption it is already familiar to candidates. Our aim is to describe the step-by-step process used in solving these problems, so that when confronted with similar problems, one should be able to apply the relevant principles.

Candidates are encouraged to solve each problem before reviewing the suggested solution and explanation that follow. By so doing, you will simulate the real exam experience and identify those areas that may require additional study. By identifying a deficiency at this stage, you will have time to do something about it before exam day. If you are troubled by the subject of parking, for example, you may investigate circulation patterns and practice drawing parking layouts until you become more familiar and comfortable with the subject.

There is no better way to prepare for the Site Planning division than solving exercise vignettes. However, the use of any single course cannot assure a passing grade. Therefore, candidates are advised to review other Kaplan products available for this division and to reference the NCARB guidelines for additional study materials.

PROBLEM 1 DEVELOPMENT RESTRICTIONS

Given the site plan shown with property line dimensions and existing underground utility line:

Assignment:

1. Using dashed lines, draw the following:
    - Front yard setback—40 feet
    - Side yard setbacks—15% of property width
    - Rear yard setback—30 feet
    - Service easement—40 feet centered on rear property line
    - Utility easement—25 feet from north property line

2. Label and dimension all setbacks and easements.

3. Outline and hatch the resulting buildable area.

Lesson Ten: Exercise Problems

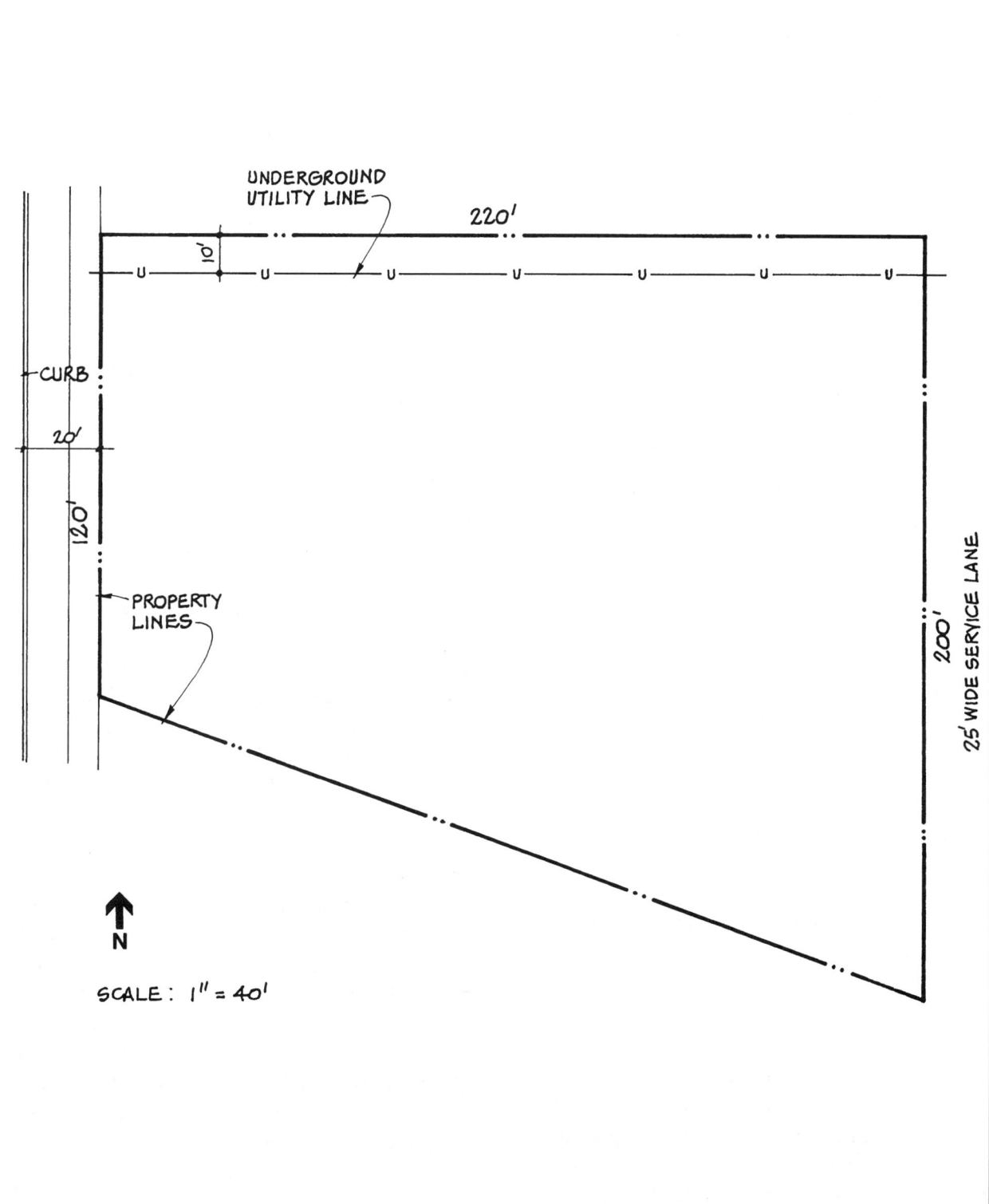

Problem 1: Discussion of Solution

Problems involving development restrictions have appeared on many past Site Planning tests, and almost all of them follow a similar form. Candidates are presented a specific site with established boundaries, and the development of this site is limited by a variety of setbacks, easements, and other such restrictions. The first concern is understanding the written description of each restriction. The next problem is to correctly interpret each restriction graphically. Finally, the last problem is to recognize those restrictions that are most limiting and which result in the largest legitimate buildable area.

In solving the problem presented here, you must proceed one step at a time so that no restriction is overlooked. For example, our first restriction states that the front yard setback is 40 feet. Since there is a curb at the west, which implies that a street exists beyond, the front of the lot is at the west. We measure the 40-foot distance from the property line (never the curb or street) and draw a dashed line that parallels the front property line. Incidentally, all dimensions are measured 90 degrees to property lines.

Next, we are told that side yard setbacks are 15 percent of the lot width, but in this case, the lot width varies from 120 feet at the street to 200 feet at the rear. Therefore, we must compute our setback at both the west and east ends and connect these with straight lines. At the west, 15 percent of 120 feet is 18 feet, and at the east, 15 percent of 200 feet is 30 feet. We indicate these dimensions at both the north and south sides of the site and draw our straight connecting dashed lines. Incidentally, it is incorrect to use 15 percent of the average of the front and rear property line dimensions [15% × (120 + 200) ÷ 2 = 24 feet], since that will produce an accurate side yard setback only at the midpoint of those lines.

Our next restriction is the rear yard setback of 30 feet, which we measure from the rear property line and draw as before. We are told that a 40-foot service easement is centered on the rear property line. Therefore, one-half of that easement (20 feet) is measured on each side of the rear property line. Since we already have a 30-foot setback at the rear, this 20-foot restriction is less limiting, and therefore, not applicable in defining the buildable area.

Finally, we must draw the northern utility easement, which is parallel to the north property line and 25 feet from it. Once this easement is drawn, we can see the peculiar situation that results. At the northwest, the utility easement of 25 feet governs, and at the northeast, the side yard setback of varying width is more limiting. Thus, it takes a portion of both limit lines to describe the northern boundary of the buildable area.

With the four building limit lines drawn, we have now represented the maximum buildable area that complies with all the restrictions presented. All that is left to do is outline and hatch this area and label and dimension all setbacks and easements.

Meanwhile, what about the underground utility line that was located on the original site plan 10 feet from the northern property line? As it turns out, this is an unnecessary piece of information that has no effect on the size of the buildable area. Problems such as this may contain data that have no relevance to the solution. Nevertheless, candidates are obliged to review every bit of information before discarding those that may be superfluous.

Lesson Ten: Exercise Problems

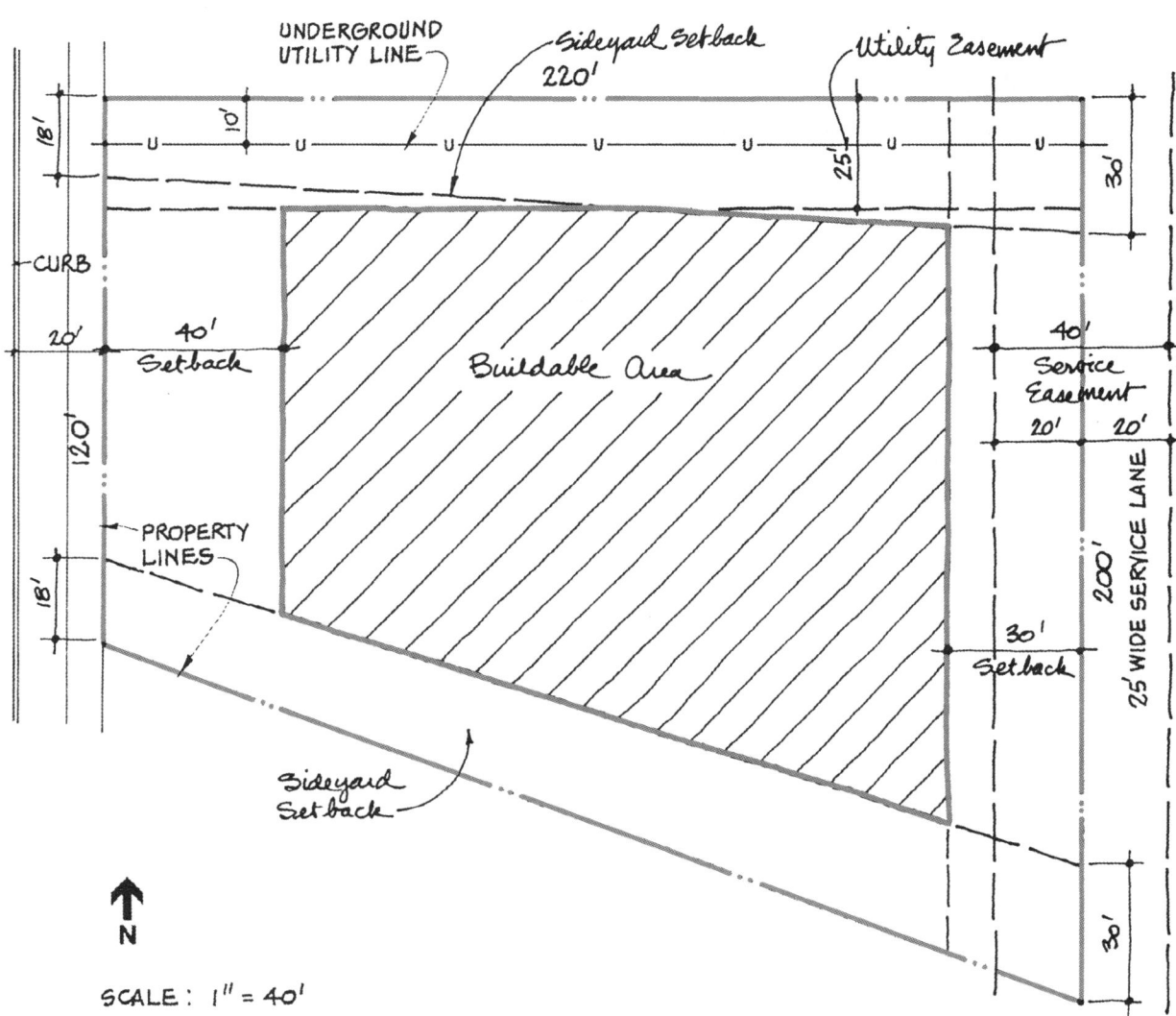

PROBLEM 2 TOPOGRAPHY AND DRAINAGE

Given the site plan shown with the location of a 60' × 125' building pad indicated:

Assignment:

Modify the existing contours so that the building pad will be at a constant elevation of 10.5 feet. The following restrictions apply:

- The slope of land within the property lines may not exceed 10 percent.
- Surface water should be drained to the site corners.
- Grading beyond the property lines is not permitted.
- Indicate revised contours with solid lines.

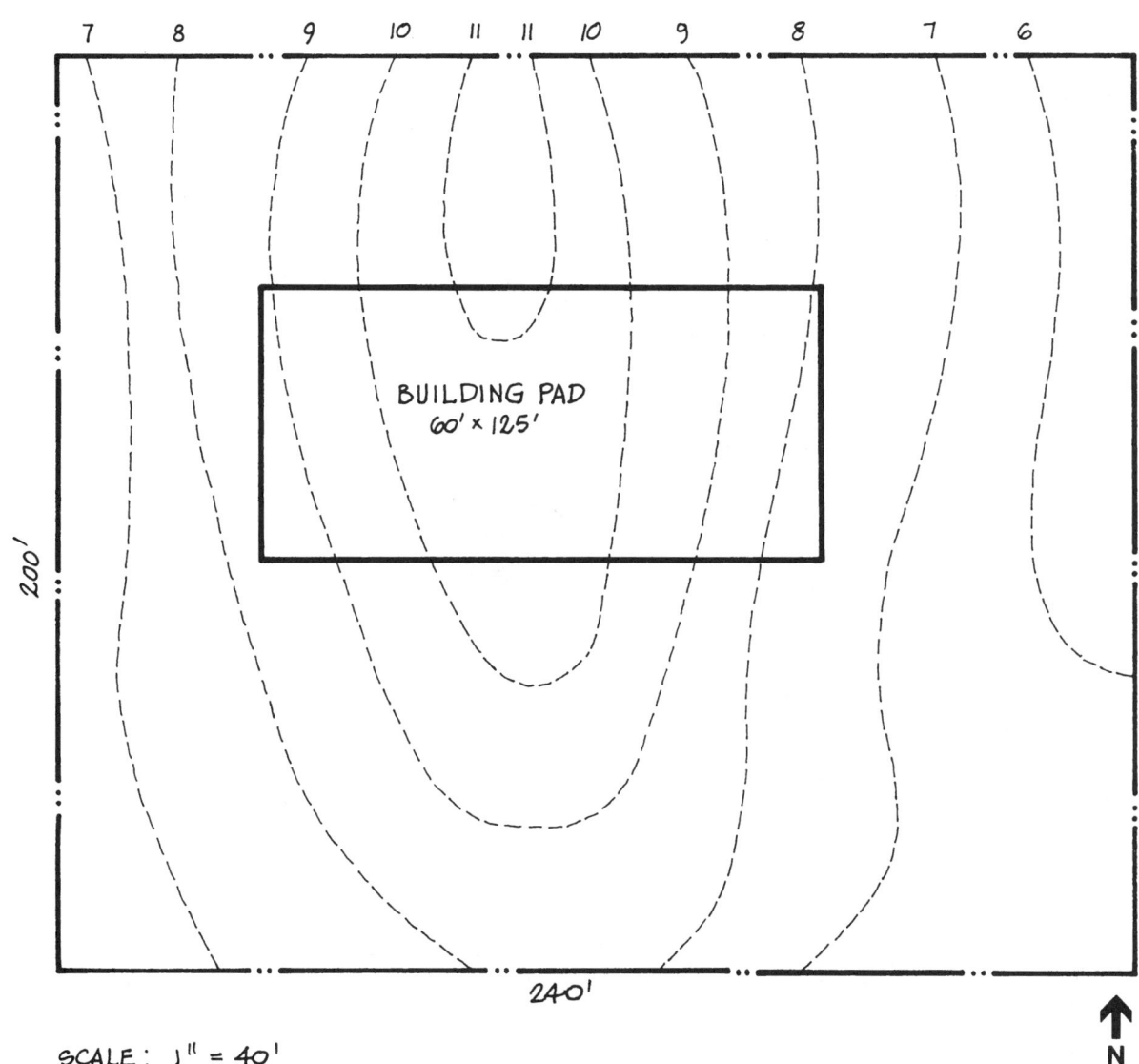

Problem 2: Discussion of Solution

This is actually two distinct problems in one, in which the topography must be modified before considering the drainage. The reason we solve this kind of problem in this sequence is in order to observe the final land shape before creating the necessary drainage swales that cut through the finished contours. Before any work is done, however, candidates should study the existing topography to determine the natural land forms.

We see a ridge-like configuration of land that has its summit at the center and then falls away toward lower elevations at the east, south, and west. The summit has an elevation slightly above 11 feet, and we are told that the graded building pad will cut through this summit at an elevation of 10.5 feet.

To avoid confusion, each affected contour should be modified in order, one at a time. We begin with contour 10, which is the one just below the finished building pad. We know that this contour lies exactly one-half foot below the pad (10.5' − 10' = 0.5'), but at what distance from the pad should it be drawn? The restriction states that the maximum land slope is 10 percent. Therefore, if we recall that the gradient (G) equals the vertical dimension (V) divided by the horizontal dimension (H), we can easily solve for H. That is, if G = V/H, then H = V/G, or H = 0.5' ÷ 10 percent = 5 feet. Thus, contour 10 is placed 5 feet away from the building pad on all four sides, which represents a 6-inch drop in elevation all around the pad.

Using the same kind of computation, we determine that all the remaining one-foot contours should be drawn exactly 10 feet apart, since one foot vertically divided by 10 feet horizontally equals our required gradient of 10 percent. Each modified contour line is drawn parallel to each side of the pad and reconnected where the revised contour meets its existing counterpart.

As one can see, every contour line is affected, except contour 6 at the east, which is already located more than 10 feet away from modified contour 7, and therefore requires no further revision. When completed, we can see on all sides a series of straight, parallel contours, evenly spaced, which is the topographic representation of a plane surface. And in fact, that is what we have created—a plane of earth on each side of the pad that slopes down and away, similar to a truncated pyramid.

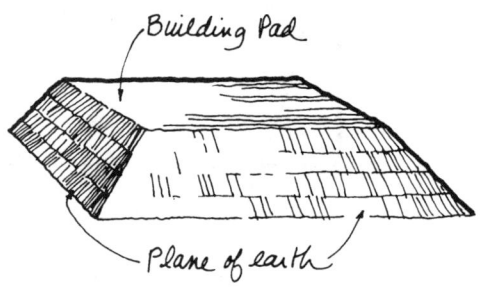

To solve the drainage part of the problem, we must create a drainage swale for each of the four corners of the site, towards which the surface water will flow. As an example, to create the swale in the southwesterly quadrant, we project finished contour 10 slightly beyond existing contour 10, make a loop, and return the modified contour to its existing counterpart, as shown below.

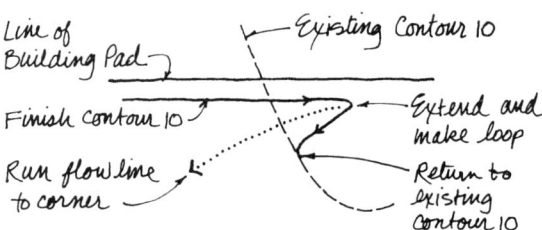

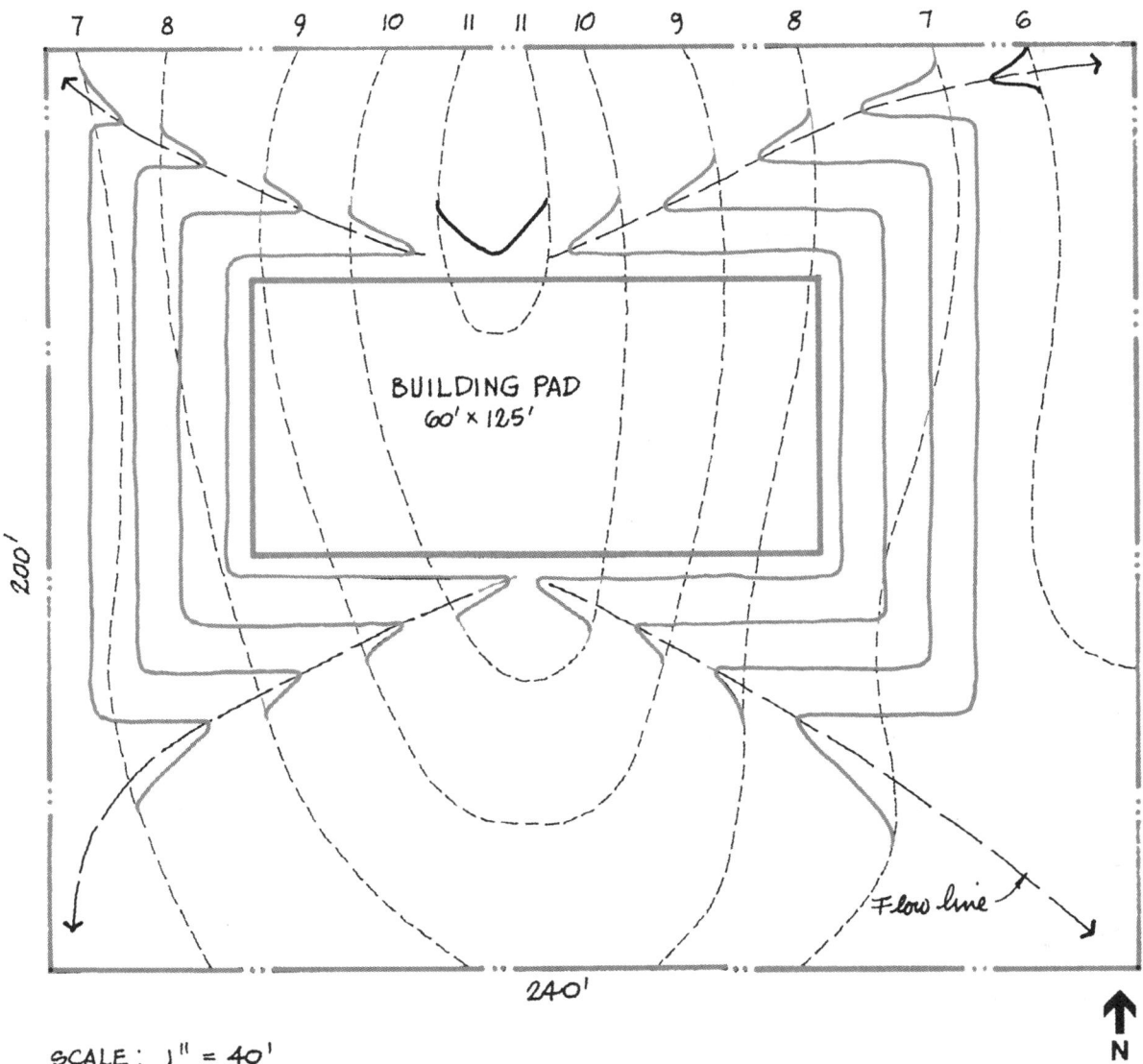

Using this loop as a starting point for water flow and the southwest corner of the site for the termination point, we draw a flow line connecting the two points. Each finished contour line is then extended and looped around the flow line in a similar way.

In designing a drainage system that utilizes swales, there are a couple of problems that must be avoided. First, no part of any revised contour may come closer to an adjacent contour than 10 feet. In other words, the 10 percent slope restriction applies to all parts of every contour line. Next, the revised contour loops should be spaced approximately equally in order to produce a uniform flow of surface water. Candidates may recall that closely spaced contours produce swift-flowing water, while more widely spaced contours produce slow-moving water. Although the actual speed of flow is rarely important in problems such as this, one should strive to create a flow path that allows water to run at a uniform speed.

Note also that the revised contour at elevation 6 in the northeast corner terminates where its corresponding existing contour line meets the property line. It is not permissible to have the modified and existing contours merge at any point beyond the property line, because if they did, you would be grading your neighbor's property and violating an important condition of the problem.

Finally, when all the swales have been completed, candidates must remember to revise contour 11. We might have added a couple of small loops here, as shown to the right.

However, we have chosen instead to develop a small ridge (contours pointing downhill). This configuration assures that water in this area will flow either eastward or westward, towards the new drainage swales. With every contour line now accounted for, our assignment is

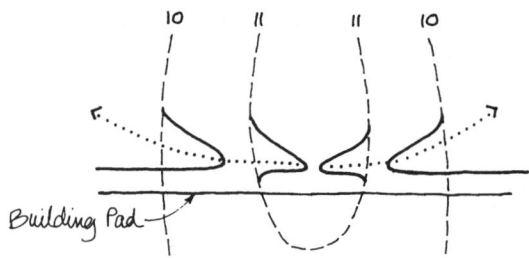

complete. Although one of our assignments was to drain surface water to the property corners, candidates may notice that quite a bit of water at the east and west drains toward the side lines. Given the problem conditions, a certain amount of this is unavoidable. However, if one had the time and the inclination, this solution could be made even more complete by adding two more swales at the east and a slight ridge at the west. Both of these improved conditions are illustrated below.

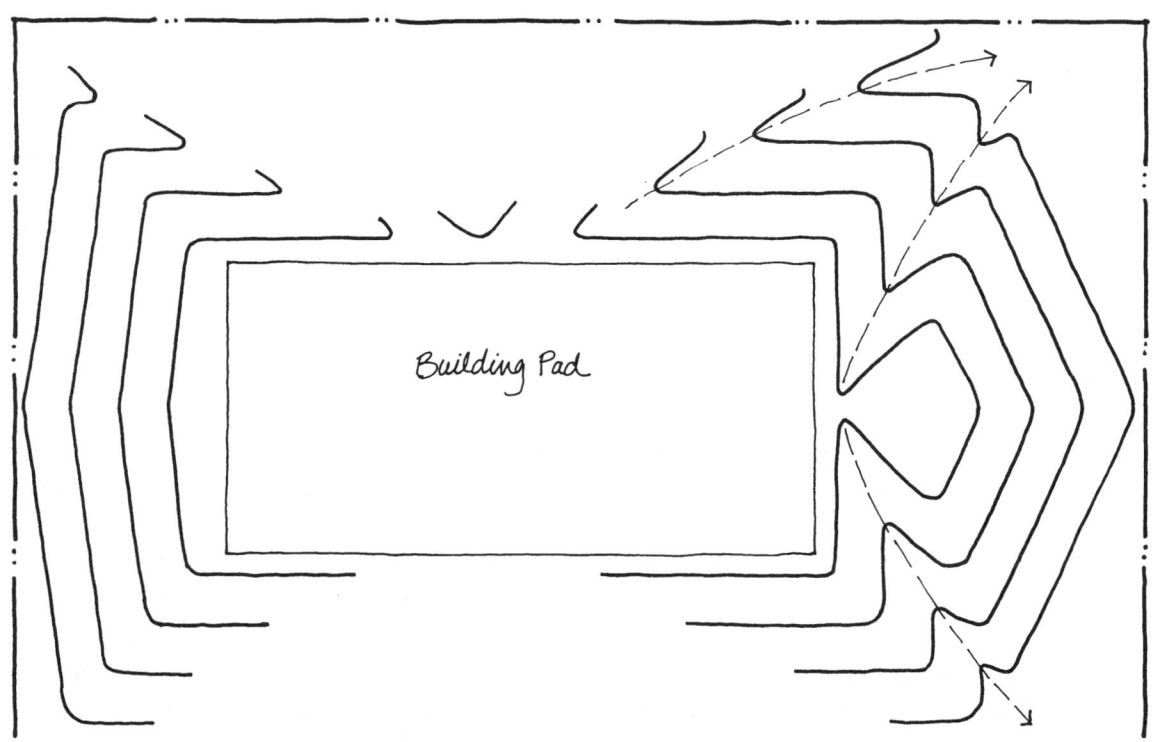

PROBLEM 3 BUILDING PLACEMENT

A local club wishes to accommodate a club house, swimming pool, tennis court, bleachers, and a service building on the site shown. The scale of all elements is indicated, and to those elements shown must be added 10-foot-wide (minimum) access walks connecting all elements, and a sun terrace of 2,000 square feet, which is adjacent to the pool.

Assignment:

Prepare a sketch plan considering the following objectives:

- All trees must be preserved.
- Bleachers shall serve the tennis court.
- The pool shall be adjacent to the club house.
- The service building shall relate to the club house and parking lot.
- The tennis court shall be oriented in the north-south direction.

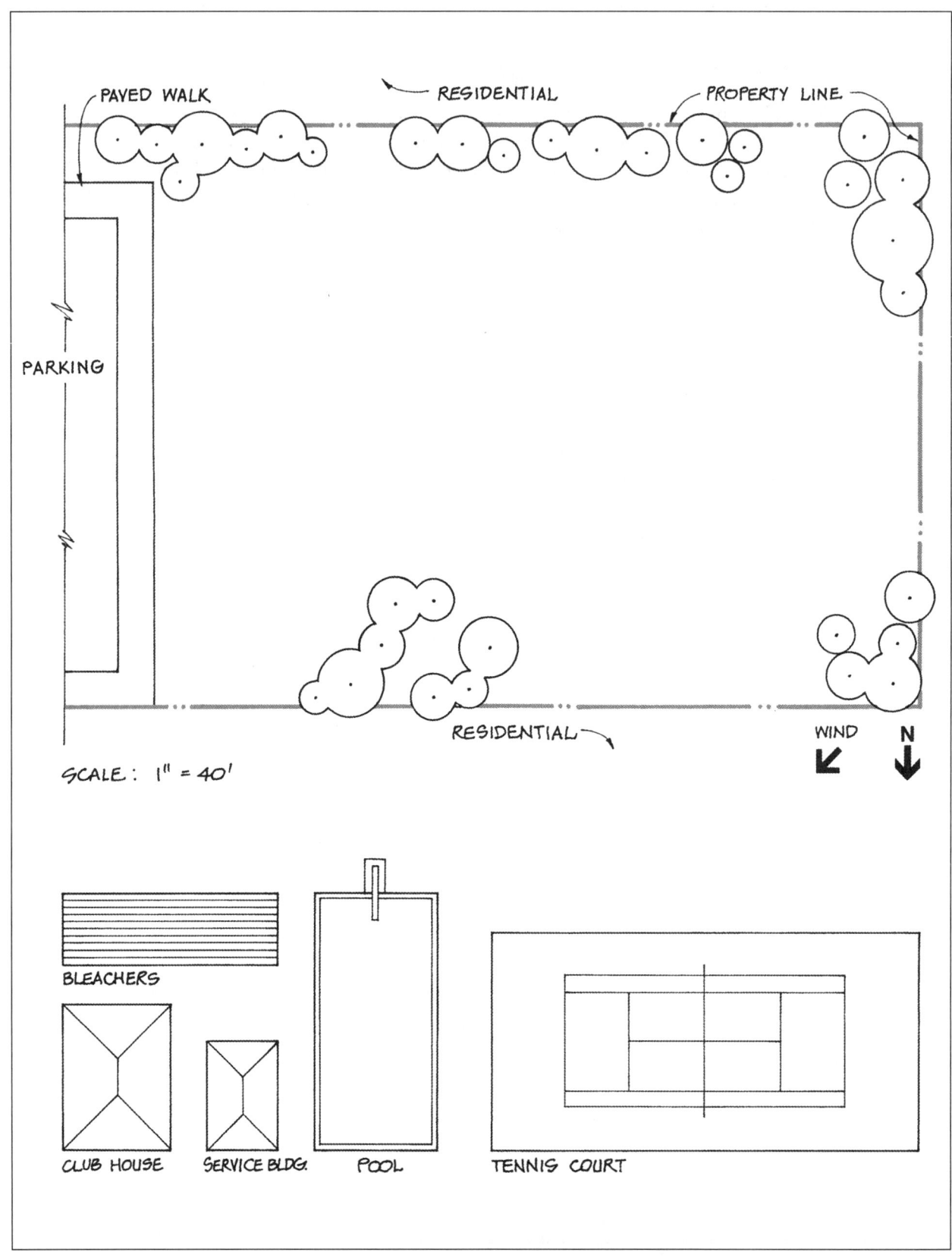

Problem 3: Discussion of Solution

Nearly every Site Planning test contains some element of building placement, in one type of problem or another. Placing a building on a site is one of the most elementary exercises presented to candidates, and without doubt, one of the most important concerns facing architects on a regular basis. Siting a building embodies fundamental design in its most essential form. One must consider function, climate, topography, context, aesthetics, and many other equally significant influences.

When a building is poorly sited, the users may suffer for as long as the building exists. On the other hand, when a building is skillfully sited, such as Frank Lloyd Wright's masterpiece, Fallingwater, both the site and the building are enhanced, and the users are enriched every day of their lives. The solution to the small problem presented here will not lead to either extreme result. However, it will provide good practice for this popular type of exercise.

To begin this problem, candidates should review both the site conditions and the building elements which must be placed on the site. We see a large open space, set in a residential neighborhood, with a parking area indicated at the east. The existing trees are almost continuous along the south property line, and there are two more groups of trees located on the north side. Since there are no contours indicated, we should assume the site is approximately level.

A review of the building elements indicates that the tennis court will occupy the largest area. Next in size is the swimming pool, and the buildings and bleachers are the smallest elements. Candidates must not forget that, in addition to the five elements shown, they must also provide an open sun terrace of 2,000 square feet. Thus, there are actually six elements to arrange.

All the clues necessary to solve this planning problem are contained within the problem statement. Nevertheless, candidates may still wonder where to begin. It is always best to proceed from what is already known. For example, the location of the parking lot is established, and we also know that the service building must relate to this parking lot. Therefore, the service building will be located at the east side, close to the paved walk bordering the parking lot. But where along that side should the service building be placed?

Our reasoning would develop something like this:

1. The service building is the one element that is rarely used by visitors.
2. Because of its relative unimportance to club visitors, it should be placed at a secondary location.
3. Due to the possibility of unpleasant odors (trash, pool chemicals, etc.) it should be placed downwind.
4. Owing to the nature of its function (trash, storage, equipment, etc.), it is considered to be not quite "respectable."
5. Since this is a service-oriented structure, it should probably be hidden from view.

In consideration of all the reasons above, we place the service building at the northeast corner of the site, close to the parking lot, downwind from the other elements, and partially screened by the existing group of trees.

Another objective of the program is that the service building should relate to the club house. Therefore, we place the club house just south of the service building and close to the paved walk bordering the parking lot. At this location the club house is more or less centered on the parking lot and in a position to serve as an entrance gateway to the other club facilities.

Next we must place the swimming pool adjacent to the club house, as the program dictates. Our preference here is to run the pool axis in an east-west direction and place it at the south end of the site, where the existing trees can help obstruct the prevailing winds.

Finally, we must locate the tennis court and the bleachers that serve it. We are told that the court must be oriented in the north-south direction. This orientation refers to the long axis of the court, and it is the conventional arrangement used in tennis, so that players can avoid facing the low-angled east and west sun. We place the tennis court at the west end of the site and center the bleachers on the long side, just west of the court. The other choice would have been to place the bleachers on the other long side, just east of the court. In that case, however, the spectators would have to face due west, which would be uncomfortable in the late afternoon.

We must still add our 2,000-square-foot sun terrace adjacent to the pool and the 10-foot-wide access walks to connect all the elements. The sun terrace is placed near the center of the site, which will be in full sun most of the day, and the walks surrounding the pool and tennis court are integrated with the terrace and club house entrance. We also add two access walks at the service building, one from the club house and the other from the parking lot, from which service vehicles would approach.

It should be clear to any candidate who has ever solved a site problem such as this that the neat relationships shown here do not fall into place automatically. It generally takes a number of trial-and-error sketches before it all works out. Nevertheless, one must realize that the placement of elements is always based on solid reasoning, which in turn is determined by the objectives of the program. Very little in the final solution is arbitrary. Rather, it is the result of interpreting the information and making logical decisions.

Lesson Ten: Exercise Problems **163**

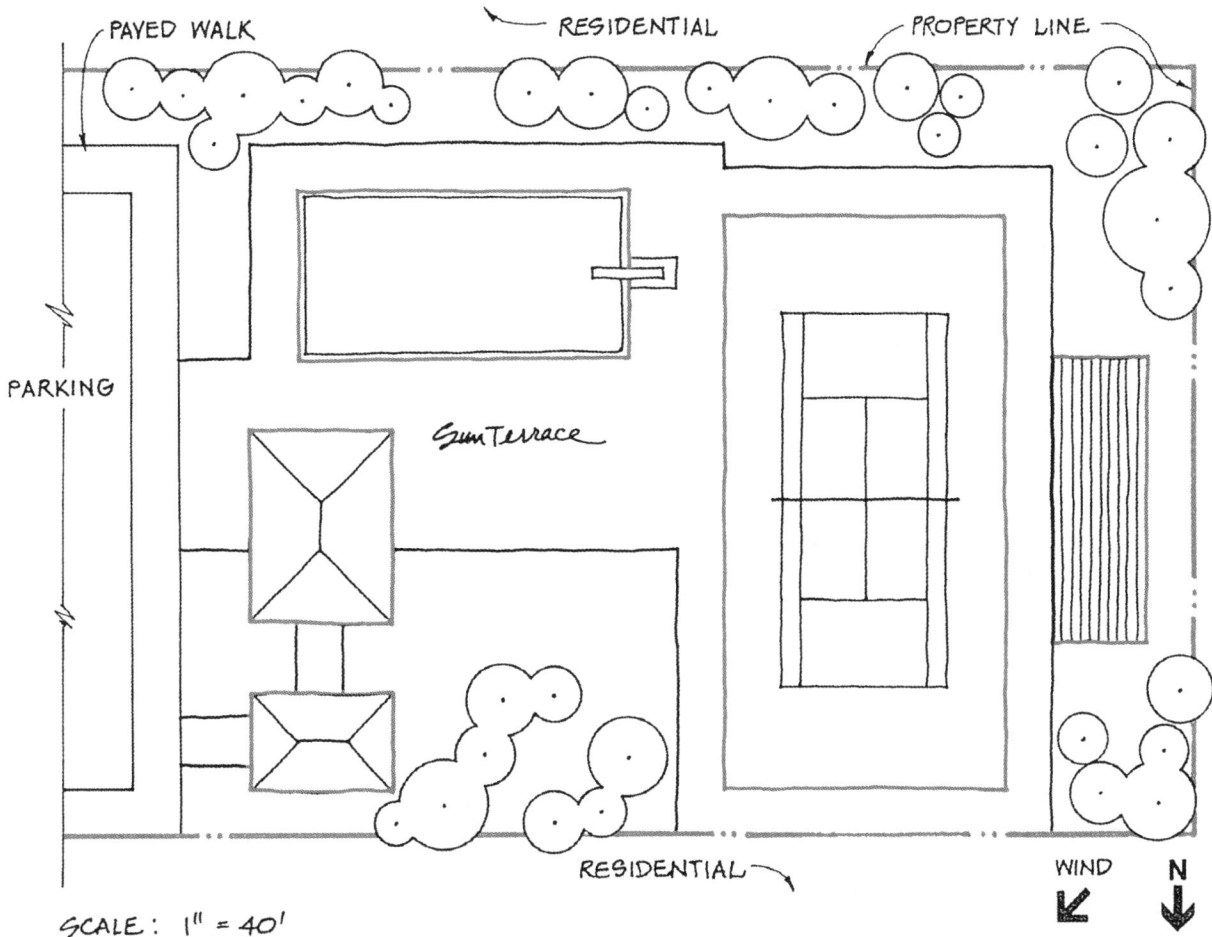

PROBLEM 4 SITE GRADING AND DRAINAGE

On the site plan shown is a house and garage, with finish floor levels indicated for each. Also shown are a paved driveway and connecting pedestrian walk.

Assignment:

Modify all existing contours necessary to accommodate the finish floor elevations indicated. All surface water should be drained around both structures and directed to the street by means of drainage swales along both sides of the site. Provide a crown in the driveway, and note all modified contours with solid lines. Maximum slopes for this project are as follows:

- Driveway—10 percent
- Pedestrian walk—5 percent
- Cut and filled earth—20 percent (except within swales)

Lesson Ten: Exercise Problems

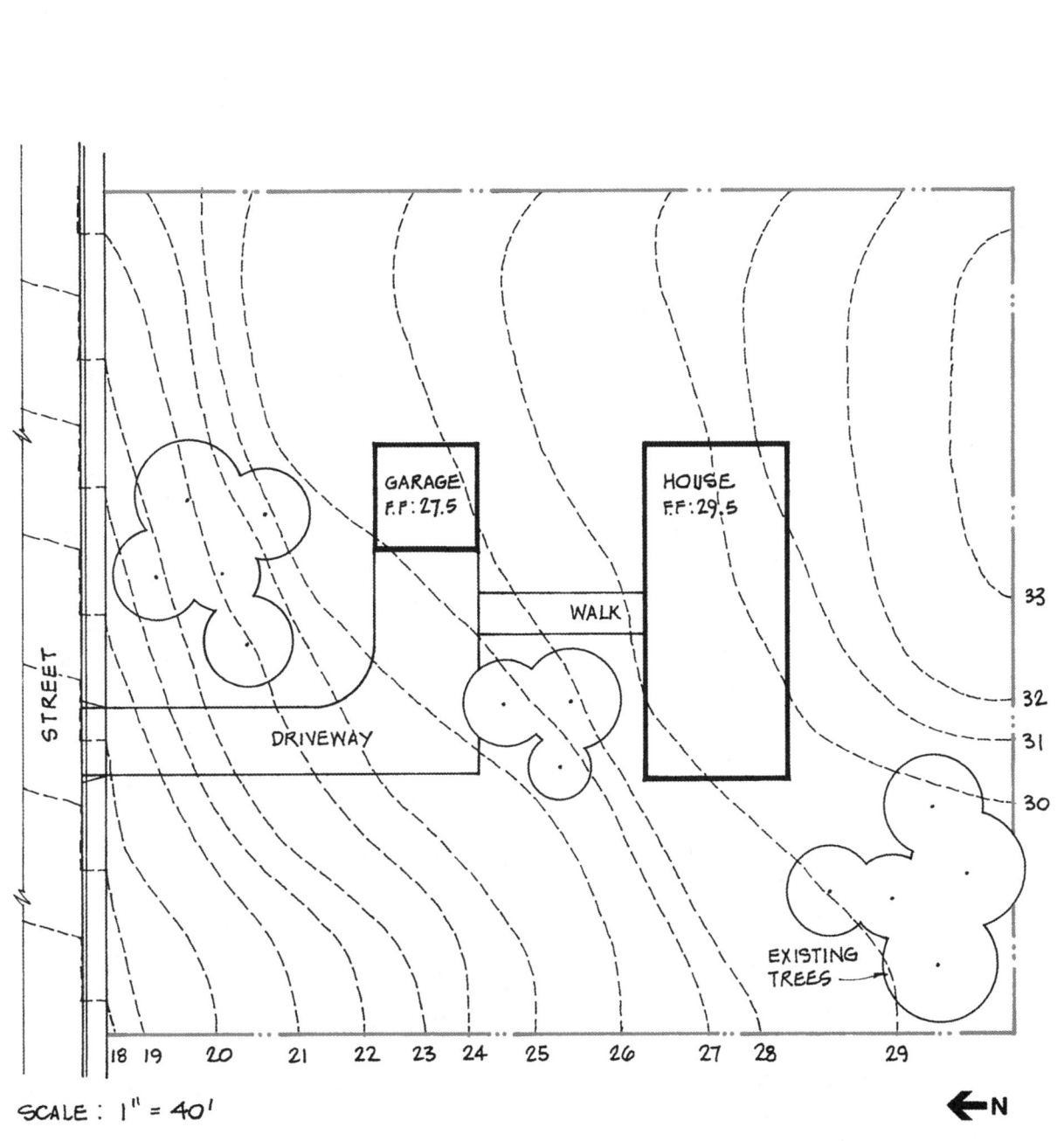

Problem 4: Discussion of Solution

Here we have another problem composed of several individual parts, and as always, the solution must be developed for each part, one at a time. Before beginning, however, one should sit back and consider what the problem is all about. First, we see a site that slopes downward from the southeast corner to the northwest corner. (Note: the north arrow points to the left!) Since the site slopes from south to north, the natural drainage is toward the street. However, our job will be to direct the surface water in such a way that it flows around the structures and follows a specific path of our own design.

The house is located towards the south, on one of the site's flattest areas, which is indicated by relatively widely spaced contours. Further north is the garage, which is also located on a relatively flat area. Toward the street we can see closely spaced contours, which indicate that this part of the site is relatively steep. Therefore, our driveway will be steep near the street and flatter as we move south towards the garage.

A good place to begin contour modification in a problem such as this is with the constructed elements, such as the paved driveway. We note that the sidewalk grade, where the driveway meets the street, is about 20. Beginning at this point on the north property line, we lay out modified contours going uphill at 10-foot intervals. One-foot contours spaced 10 feet apart result in our required driveway slope of 10 percent. By the time we reach contour 25, the spacing between contours is increased to match more closely the adjacent existing grade.

We are told that the driveway has a crown, and candidates may recall that this ridge-like configuration is indicated by rounded contours pointing downhill. Therefore, we use the points where straight contour lines intersect the driveway edges, and we draw an arc of a circle on each, pointing in the northerly direction. These arcs need not extend beyond contour 25, because at that point we are past the driveway and into a flatter area just outside the garage.

In order to keep surface water out of the garage, we must maintain a finish grade around the garage which is lower than the specified floor level of 27.5. Therefore, we take contour 27 from the south side of the garage and wrap it around the garage on the east and north, and then we bring it southward, across the paved area, to the walk, at which point it rejoins existing contour 27. By so doing, we have created a finished grade around three sides of the garage that is 6 inches lower than the garage floor.

Next, we consider the pedestrian walk between the paved area and the house. Having just established the walk elevation (27) at the north end, we must now do the same at the south end of the walk. Where the walk meets the house, we assume the finish elevation will be 6 inches below the house finish floor of 29.5. Thus, our walk has an elevation of 27 at the north, 29 at the south, and a distance of 40 feet separates the two ends. Fortunately, this works out to be a 2-foot rise in 40, or the prescribed grade of 5 percent. We can now establish the intervening contour 28 exactly midway between the two ends of the walk.

It is now time to develop the drainage swales that must run down both sides of the site. The most southerly revised contours begin just south of the house in order to redirect surface water around the structure and towards the street. As with the garage, we must maintain a finish grade outside the house which is lower than the finish floor elevation of 29.5.

Lesson Ten: Exercise Problems **167**

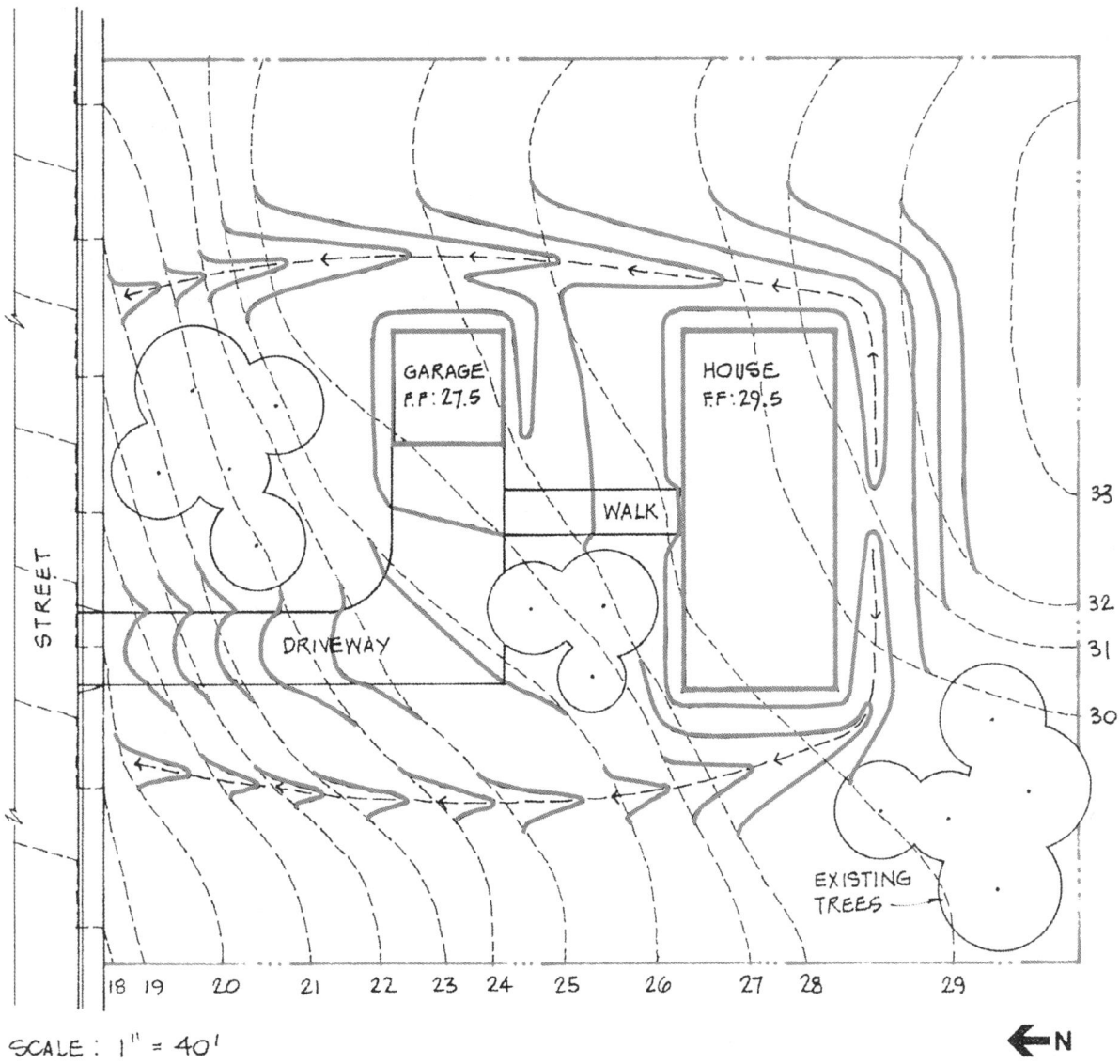

Therefore, we begin with contour 29, and we wrap this contour around all four sides of the structure. The small break on the south side allows us to create two drainage ditches, which flow in opposite directions. You may think of this as a kind of continental divide, which allows surface water from higher elevations to separate and flow either east or west.

Once the first two revised contours are established, we draw an arbitrary flow line, which represents the path that water will follow to the street. This line begins at the first contour, it continues along a smooth curve, and it ends at a culvert under the sidewalk adjacent to the street. The flow line represents a spine on which subsequent revised contours are attached, and as such, it is the pattern to follow in developing our swales.

Drawing drainage swales on a site design vignette is always a bit arbitrary, because there is rarely enough time to actually design a swale's cross section. Furthermore, this level of precision is generally unnecessary. However, swales should look "right," and certainly the contours should be spaced with some forethought. In this regard, try to space your revised contours an equal distance apart, or as shown here, with an increasingly greater distance between adjacent contours.

Our final task is to connect all the revised contour segments and be certain that all lines are reconnected to their counterparts within the property lines. That is a relatively simple process here, since there are few loose ends. However, we must not forget to modify contours 30 through 32. If you recall, the modified grade was not to exceed 20 percent. Therefore, these contours on the south side of the site must be shown at least 5 feet apart.

A note about the sidewalk culverts: Even though these were not specified, we believe it is a good idea to indicate a means by which surface water can reach the street without flowing over the sidewalk. Normally, we strongly discourage adding anything to a design program, but we believe this one exception provides an important improvement. On the other hand, those who ignored this detail would not be penalized.

Finally, candidates may have wondered about the three tree clumps on the site. Their purpose was to provide obstructions to be avoided. In other words, if a candidate ran revised contours within the tree drip lines, his or her solution would suffer. It is by no means a fatal mistake, but merely something to be aware of and avoid.

PROBLEM 5 UTILITIES PLACEMENT

On the wooded site shown is a Ranger Station, which is located directly southeast of Sierra Road. The floor of this structure is a concrete slab 6" above the adjacent grade. A stream runs through the southern area of the property, and a 50-year flood plain exists at elevation 215. The following restrictions apply to the placement of utilities:

- The septic system (tank and field) must be a minimum of 20 feet from property lines, 10 feet from structures, and 100 feet from streams.
- Both the septic tank and water well must be a minimum of 10 feet from any trees.
- Both the bottom of the septic field and the water well head must be located at least one foot above the flood plain.
- The water well must be a minimum of 50 feet from structures and 100 feet from the septic field.
- Existing vegetation must not be disturbed.

Assignment:

Using the symbols shown, locate the following utilities on the site and provide all required setback dimensions:

- Water well
- Gas line
- Underground electric service line
- Septic system (note top and bottom elevations of field)

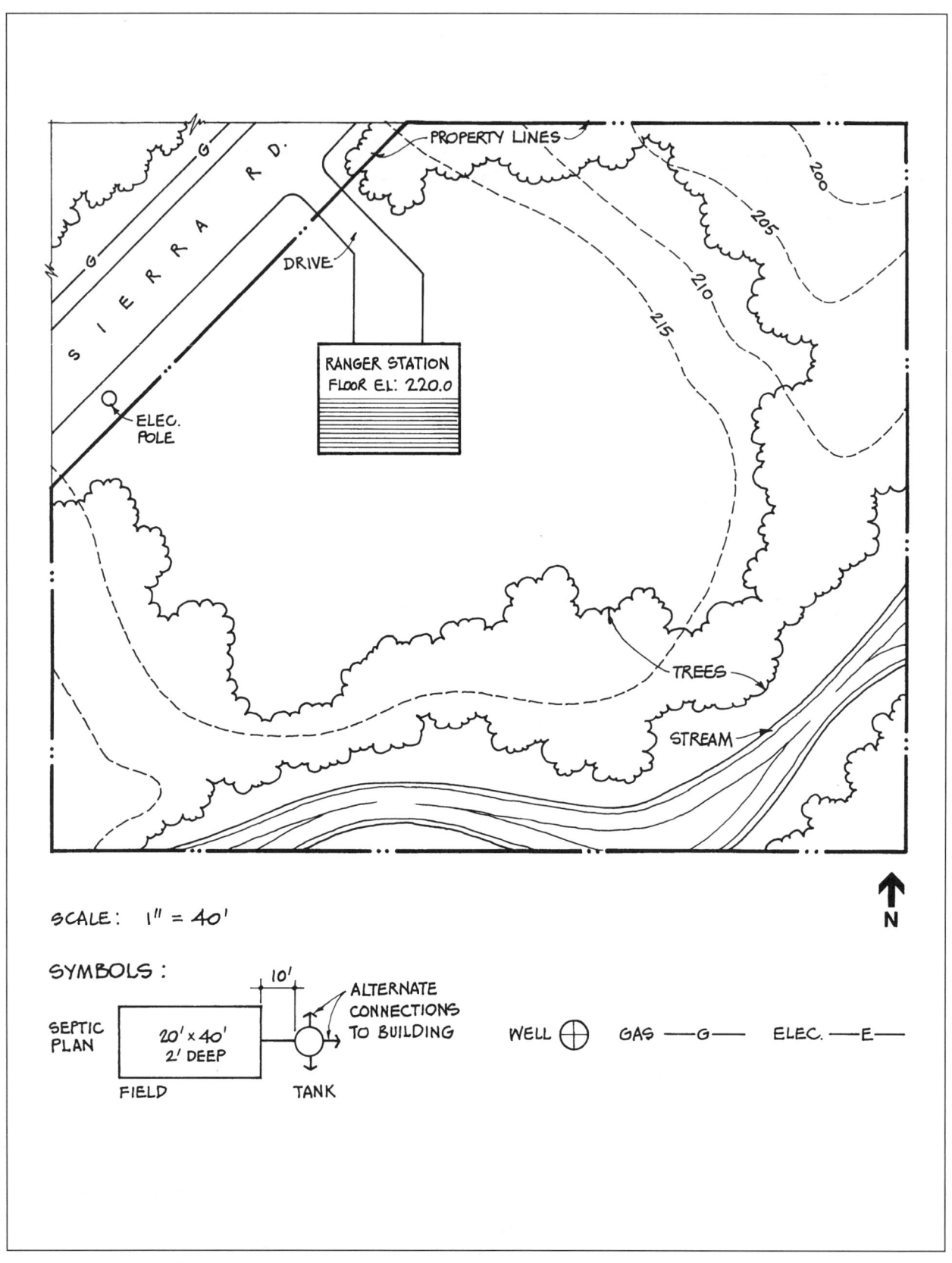

Problem 5: Discussion of Solution

This is a problem about locating utilities in a remote, natural environment. Since most of us live in developed urban areas, where problems of water and waste are generally a municipal responsibility, some discussion may be helpful. To begin with, a 50-year flood plain is an area that is expected to be flooded with water on an average of once every 50 years. In our case, this means that every 50 years or so, the volume of precipitation at our site will be so great that the stream will overflow its banks and water will rise to an elevation of 215.

Water wells are generally used to supply water in remote areas where public water sources are unavailable. Wells are bored, driven, or drilled to depths below the water table, where a reliable supply of free-flowing water is found. This source of water is generally pure, cool, and completely safe to drink. Although not a part of this problem, water well systems often include mechanical pumps and storage tanks. Regardless of a well's depth, which could be 100 feet or more, the well head (the part projecting above ground) should be located above a flood plain to avoid contamination from sediment or physical pollutants.

Septic systems are private sewage treatment systems that are commonly used where public sanitary sewer lines do not exist. The typical arrangement consists of a precast concrete septic tank, where solids settle and anaerobic digestion (occurring without oxygen) takes place, and a filtering system, where the effluent (flowing liquid part) receives a secondary treatment. Filtering devices include seepage pits, sand filters, and the system used in our problem, a drain field.

Drainage fields, sometimes referred to as disposal fields, commonly employ long lines of loose clay tiles, which are set in gravel-filled trenches. As the effluent runs through the tiles and out of the joints, it seeps slowly into the earth. Due to the waste products involved, the placement of septic elements is strictly controlled by local ordinance. And since the filtration of effluent involves leaching water into the surrounding porous soil, it is essential to locate drain fields above the flood plain, in order to avoid the possibility of dangerous contamination.

With a clearer understanding of the elements in this problem, candidates should feel more confident in devising a solution. The restrictions of this problem are so numerous that the first priority will be to locate the water well and septic system so they conform to all the required setbacks.

We begin with the septic system, because that element requires the greatest amount of space. Among the restrictions, we note that the system must be located a minimum of 100 feet from the stream. Since the distance south from the Ranger Station to the stream is less than 100 feet, it is reasonable to assume that the septic system will have to be located somewhere north or east of the building. We are restricted in both those directions by contour 215, above which the bottom of the drainage field must lie.

The precise placement of the septic system requires a considerable amount of trial and error. It is unlikely you will hit the perfect spot on your first try. However, if you sketch out setbacks 20 feet from property lines, 10 feet from the trees and building, and 100 feet from the stream, you will begin to focus on the proper area.

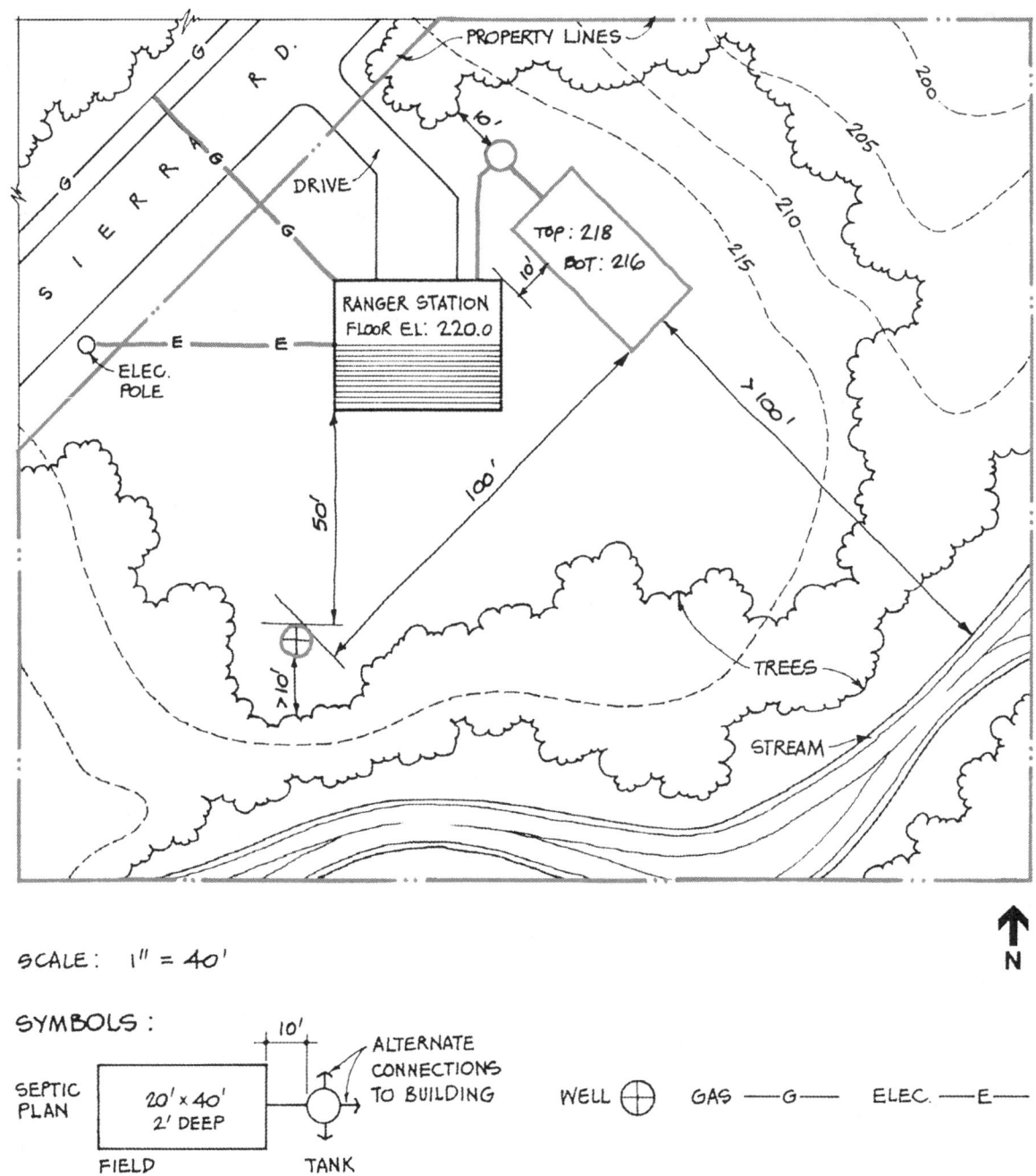

Placement of the water well is analyzed similarly. First, one must draw setback lines that are 100 feet from the septic system, 50 feet from the building, and 10 feet from the tree line. Around this target area, where the various setbacks converge, will be the right location for your well. Again, it is a process of trial and error, and you must remember this: Regardless of your initial difficulties and the confusion of all these various restrictions, every problem is designed to have a solution. So, if it does not work immediately, be assured it will work eventually.

The gas and electric lines are the simplest part of this problem. One merely draws the shortest straight-line connection between the utility and the building, and each line is identified with the symbol indicated. The gas line, incidentally, may be trenched directly under Sierra Road, and the underground electric line begins at the base of the electric pole and runs the shortest possible straight-line distance to the closest exterior wall of the Ranger Station.

Finally, we must dimension all the required setbacks and note the top and bottom elevations of the septic field. Since the bottom of the field was indicated as one foot above the flood plain, it is noted as 216, and the top, therefore, is two feet higher (see note on symbol), making it 218. Candidates should be aware that most problems of this type are designed to have one roughly correct solution. Therefore, if your solution is substantially different from this one, go back and review the list of restrictions in the problem statement. You may very well have missed something the first time around.

PROBLEM 6 SOLAR ZONING RESTRICTIONS

With reference to the three building sites shown, solar zoning restrictions prescribe that the height for all new construction shall not exceed 100 feet. In addition, the height of new construction shall not extend beyond a 45-degree plane drawn upward from the northern setback line of each site, at a point in each setback line that is 40 feet above finish grade. The following setback restrictions apply to all sites:

- Helio Street—15 feet
- Sol Avenue—20 feet
- Side yards—10 feet
- Rear yards—15 feet

Assignment:

On the graph below the site plan, draw a north-south section through the center of all three sites, which shows the maximum allowable building envelope. Dimension and hatch all resulting forms.

Lesson Ten: Exercise Problems

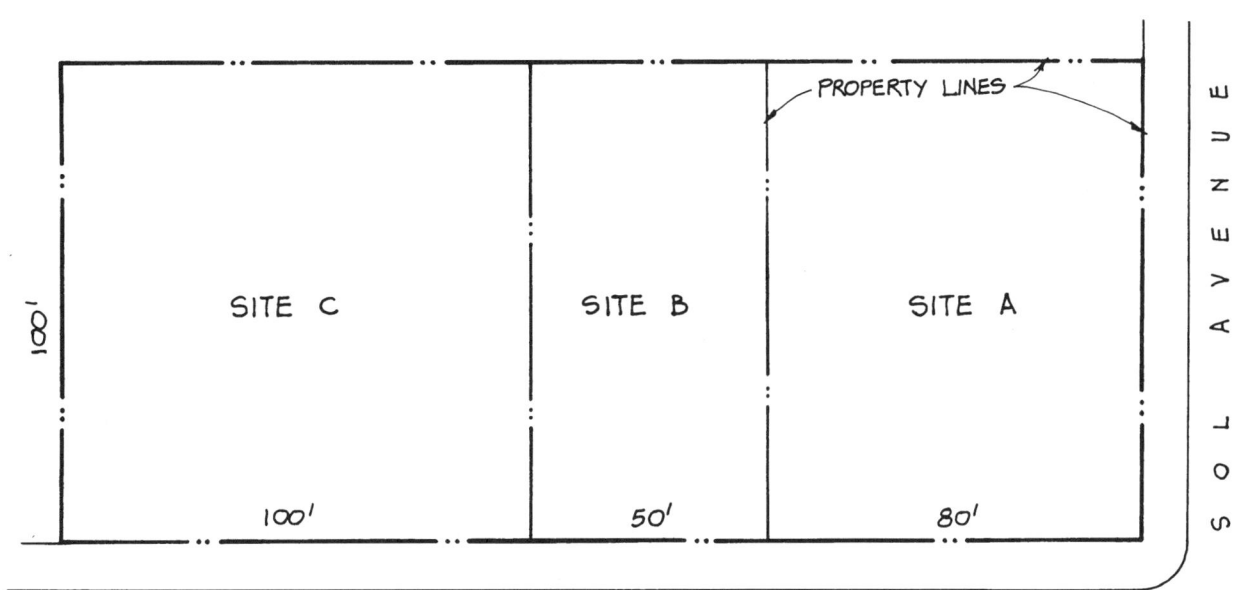

PLAN   SCALE: 1" = 40'

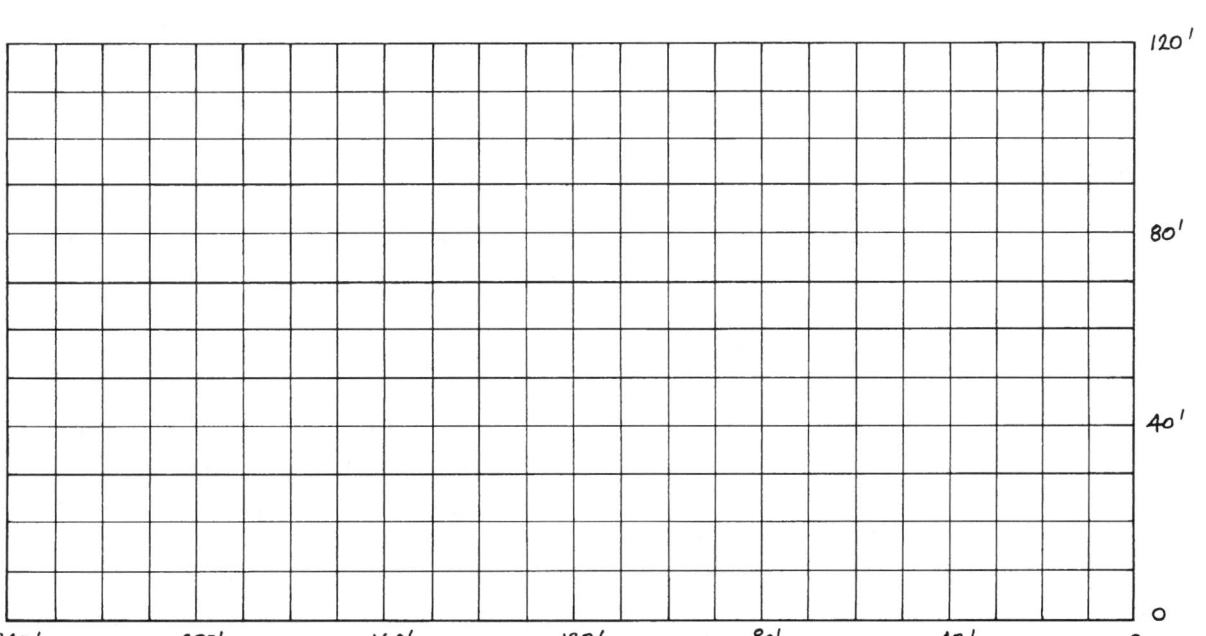

SECTION   SCALE: 1" = 40'

Problem 6: Discussion of Solution

This is another development restriction problem, representing a type of vignette that has become prevalent in recent years. In this case, the restrictions apply to both plan and section, but otherwise, it is little different from any of the others. When a problem such as this one first appeared on the exam, many candidates were completely bewildered by the mere idea of solar zoning, because so few had experienced doing such studies. Those more familiar with site design vignettes, however, recognized that this problem had little to do with anything solar. It was simply one more development restriction exercise.

As in all problems of this type, we want to define the buildable area, and in this case, it is the buildable area in section that must be established. Since the section to be drawn on the graph will cut through the site in the north-south direction, we are interested only in the north-south dimensions. Therefore, we have no interest in a dimension such as the Helio Street setback, since that will not appear in any north-south section cut through the site.

We could actually begin with any one of the three sites, but we choose to start with corner Site A at the far north. We are told that the setback along Sol Avenue is 20 feet. Referring to the section-graph, we note that the zero vertical line lies directly below the northern property line of Site A. One should also note that the plan and section are drawn to the same scale, so that plan dimensions may be projected directly downward from plan to section. Thus, we move along the bottom scale of the section to the 20-foot mark (halfway between 0' and 40'), and we draw a vertical along that line. This line represents the most northerly face of any structure that can be built on Site A.

The next limit line to identify is the southerly setback line of Site A. However, because this site is bordered by two streets, we must first determine whether the line separating Sites A and B is the rear line or the side line of Site A. If it is the rear line, the setback would be 15 feet, but if it is the side line, the setback would be only 10 feet.

Candidates should remember that, as a general rule, the short side of any corner lot is considered to be the front of the lot. Therefore, Site A fronts on Helio Street, and the 100-foot-long line between Sites A and B is a side line for both sites. Referring again to the bottom scale of the section-graph, we plot the southerly setback line for Site A, which falls 10 feet north of the site's southern property line.

We must now deal with the 45-degree plane that is described in the problem statement. The instructions prescribe that this angle begins at a point on the northern setback line that is 40 feet above grade. Therefore, we begin (on the section) with the setback line closest to Sol Avenue, and we move upward along that line to the 40-foot horizontal line, which is noted on the scale to the right. The point at which these two lines intersect is the point from which we project the 45-degree angle upward and to the left.

Now, one may ask, how do you know that the 45-degree angle is drawn up and left, rather than up and right? First of all, this angle begins at the northerly setback, and to complete the figure, it must connect to the southerly setback line. But more important is the answer revealed by the section itself. Since the purpose of solar zoning is to permit sunlight to penetrate more deeply between buildings, one can see that the southerly sun will cast its rays down and to the right. If the angle on the envelope were reversed, the buildings would block the sunlight

Lesson Ten: Exercise Problems

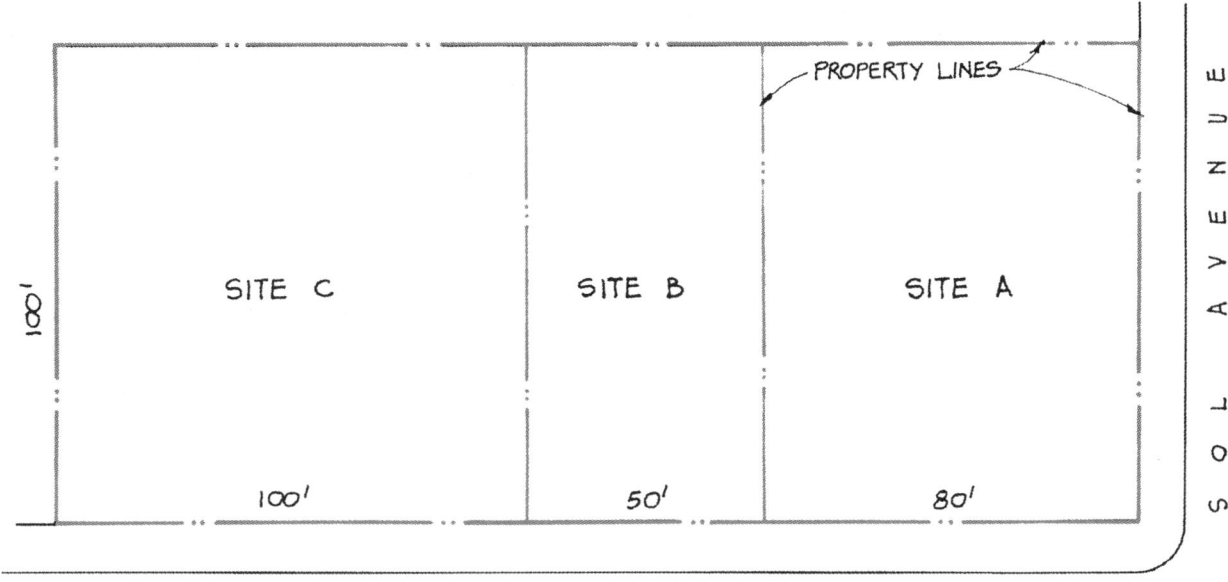

PLAN   SCALE: 1" = 40'

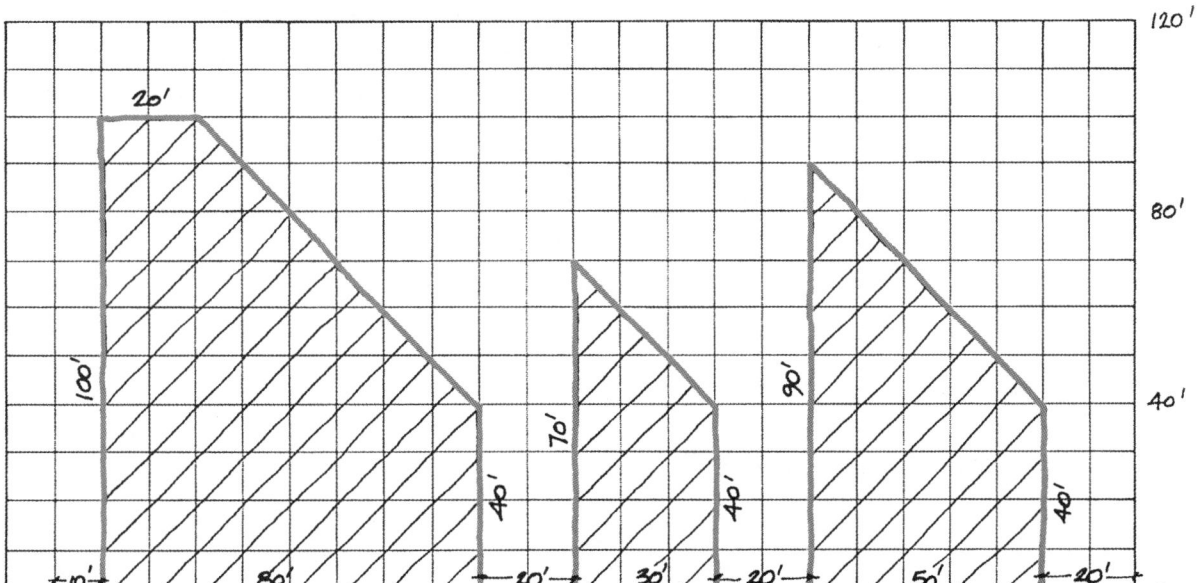

SECTION   SCALE: 1" = 40'

at the side yards, instead of allowing it to pass through.

As one can see, the 45-degree angle intersects the southerly setback line of Site A at exactly elevation 90, as measured along the vertical scale at the right of the graph. The completed figure represents the maximum allowable building envelope for Site A, or, put another way, the section of any structure built on Site A must fall within the shape just created.

Moving on to Site B, we analyze the restrictions in the same way. Here we have a 50-foot-wide frontage on Helio Street, and there are side yard setbacks of 10 feet at both the north and south sides. We project this data directly downward to the graph-section and show the two vertical limit lines at 90 and 120 feet, as measured along the bottom scale of the graph. Once again we project a 45-degree angle that begins at the intersection of the north setback line with the horizontal 40-foot height line. And once again this angle moves up and to the left until it intersects the southerly setback line of Site B. This completes the building envelope for Site B.

The shape of the building envelope for Site C is constructed in the same way as previously described. However, in this case the 100-foot maximum height limit comes into play. Therefore, even though the 45-degree solar angle would intersect Site C's southerly setback line at elevation 120, the top of the envelope is cut off at the 100-foot-high limit line, which produces a 20-foot-wide flat top on the resulting shape. With the completion of the allowable building envelopes on all three sites, we have only to hatch the final forms and provide all key dimensions to complete our solution.

At this point, candidates should understand the significance of solar zoning. As the sun moves across the southerly sky, the shape of buildings within their respective envelopes allows sunlight to penetrate between buildings. Without the required 45-degree plane, a structure on each site could rise to 100 feet, and the result would be dark, narrow canyons between buildings. Therefore, solar zoning is an example of a municipal ordinance used to control and preserve the urban environment for the benefit of all.

PROBLEM 7 DEVELOPMENT RESTRICTIONS

The site shown has the following development restrictions:

- No building is permitted within 50 feet of the cliff edge, except that building within 50 feet of the cliff edge is permitted above elevation 25.
- No building is permitted below elevation 20.
- No building is permitted closer than 20 feet to the drainage channel centerline.

Assignment:

Show by outline and hatching all buildable land area.

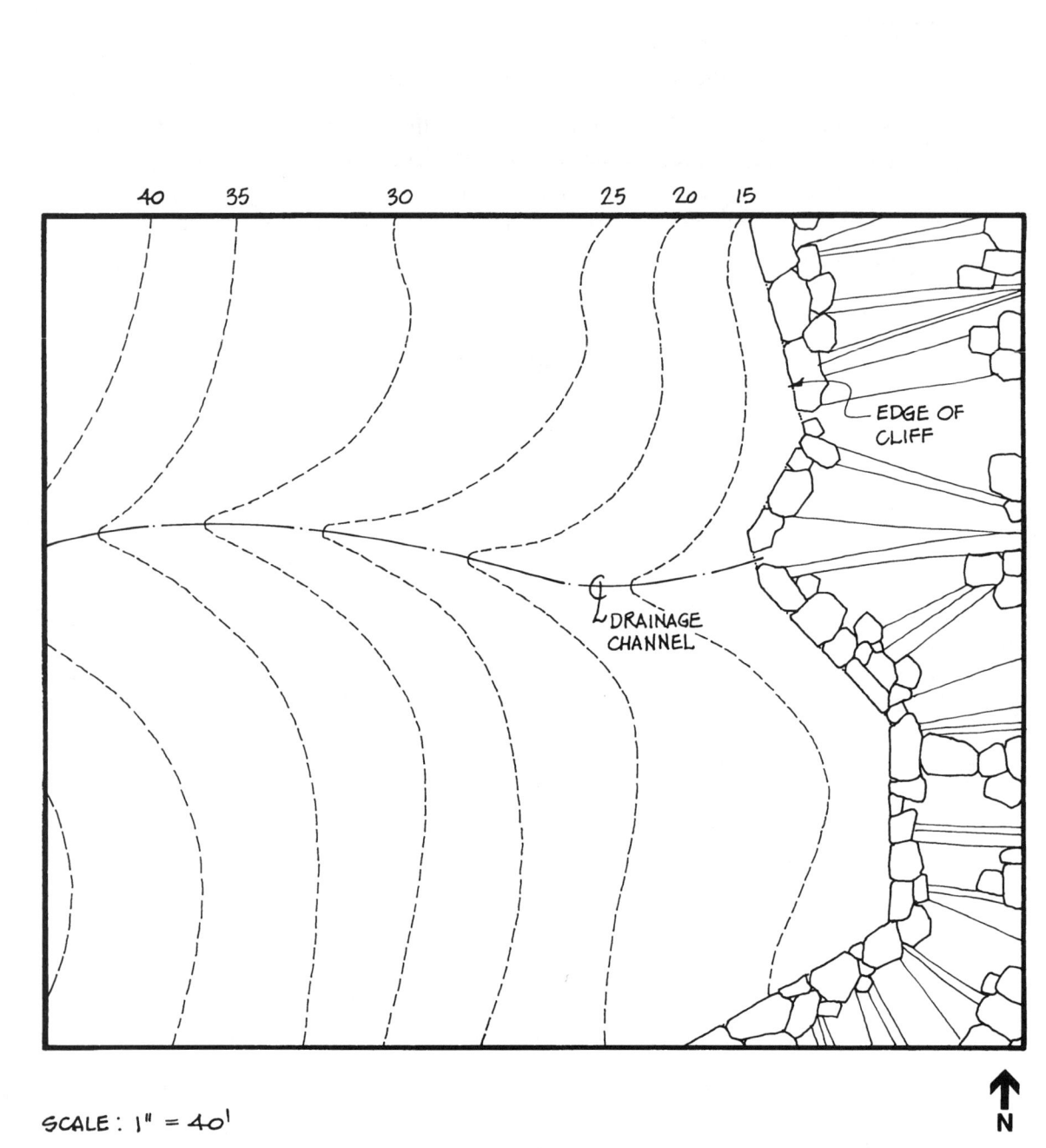

Problem 7: Discussion of Solution

Most development restriction problems appearing on the Site Planning exam deal with setbacks and easements on conventional building sites. As such, one is generally faced with rectangular lots and restrictions that follow straight lines. In this problem, our restrictions follow a rambling cliff edge, a curving drainage channel, and irregular natural contours. Nevertheless, the principles governing restrictions of development remain the same, and as you will soon see, this is really a fairly simple problem.

As in all problems of this type, we are trying to determine the shape of an area on which building is permitted. Put another way, we are seeking land that is unaffected by any of the restrictions contained in the program statement. Therefore, the restrictions in the program must be considered separately, and they must be delineated one at a time.

We begin with the first restriction, which states that no development may occur any closer to the cliff edge than 50 feet. Therefore, we must draw a setback line that is 50 feet from, and parallel to, each segment of the cliff. One can generally scale a single 50-foot dimension (always measured perpendicular to the cliff edge) and then estimate or "eyeball" a line parallel to each segment. However, for those interested in greater accuracy, scaling two points along each line will result in a line with more precision.

Next, we are told that above elevation 25, building is permitted nearer to the cliff than 50 feet. Therefore, at the north, where contour 25 crosses to the east of the 50-foot setback line, the existing contour becomes the limit line for development. In other words, the 50-foot setback does not apply at this point, because this new restriction is more lenient. In fact, we can build as close as we want to the cliff edge, providing the natural grade has an elevation of at least 25. The same restriction applies at the south, where contour 25 crosses to the east of the 50-foot setback line.

The next restriction states that no building is permitted below contour 20. This means that no matter how far you are from the cliff's edge, even if it were a hundred feet or more, you simply cannot build on any land where the existing grade is lower than 20. This restriction applies here south of the drainage channel, where three segments of the 50-foot setback line intersect, east of contour 20. Since this new restriction is more limiting, the buildable area must be cut back to contour 20.

The resulting east side setback line is now complete, and the effect of all these restrictions is a building limit line made up of segments of the 50-foot setback line, as well as portions of existing contour lines 20 and 25. It is important to recognize that this line must be continuous, regardless of the number or shape of individual segments.

The remaining restriction requires us to draw a 20-foot setback line on each side of the drainage channel, since no development is permitted closer than 20 feet to the channel centerline. Again, this may be plotted freehand, using a few measured points on each side of the centerline. These two new setback lines produce a 40-foot-wide strip of land that straddles the drainage channel and intersects the east side setback line 50 feet from the edge of the cliff.

The buildable area resulting from all the restrictions is a small, nearly rectangular piece of land at the north and a larger, odd-shaped piece of land at the southwest. Because of the existing drainage channel, one should have known from

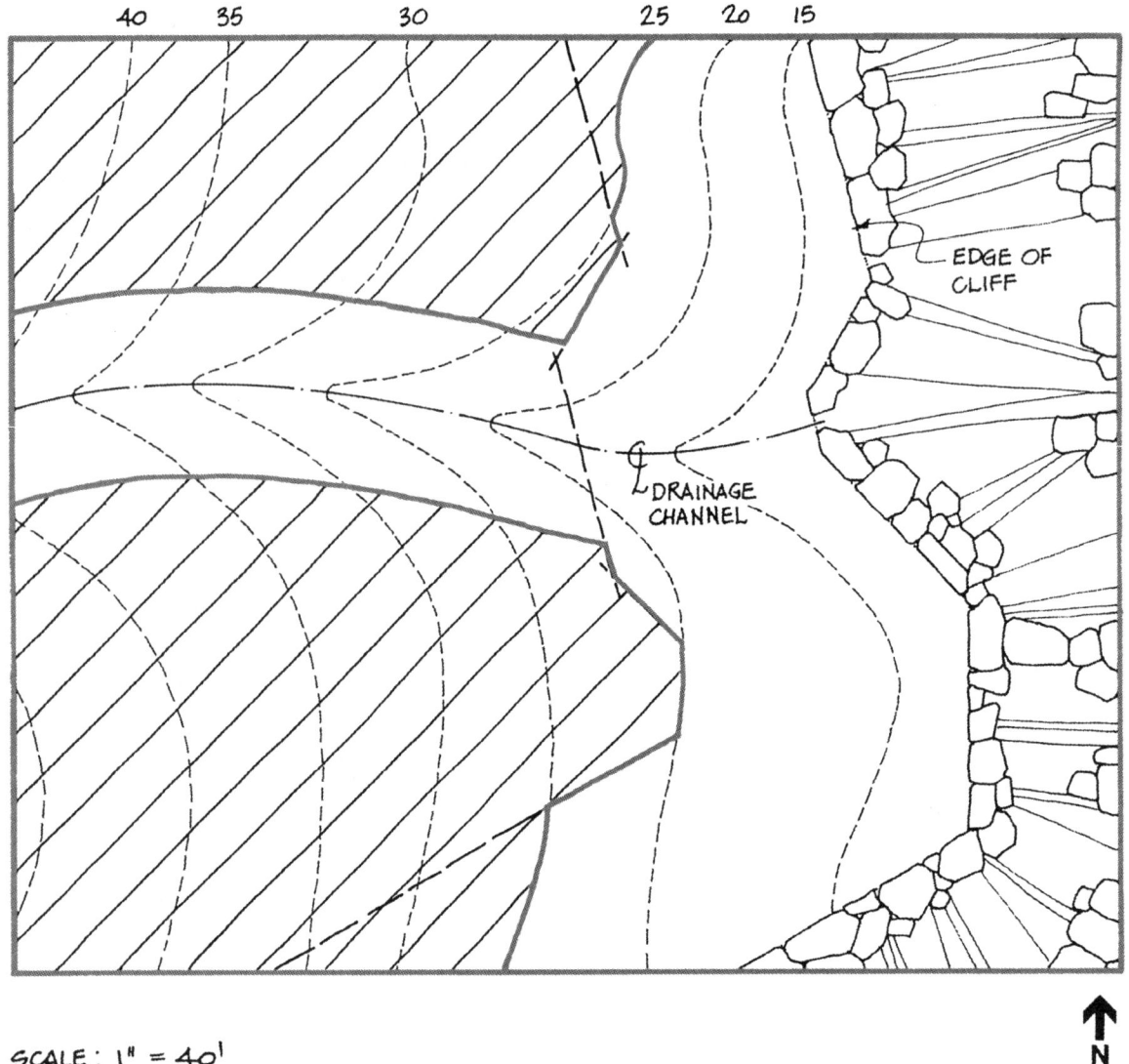

SCALE: 1" = 40'

the start that the buildable land would comprise at least two separate areas. When these two areas are outlined and hatched, our solution is complete.

We mentioned earlier that this problem was fairly simple. Whether or not you came up with an error-free solution in 10 minutes (a reasonable average), you must admit that there is little about this problem that is complicated. Those who had trouble may have forgotten the most important principle: The different parts of the problem must be considered and solved one at a time. If you remember that strategy, you will have little trouble with problems of this type.

PROBLEM 8 SITE GRADING

Given the property shown with existing 2-foot contours indicated.

Assignment:

Modify the existing contours to create a level rectangular area 80 feet by 120 feet in size. The level pad may be located anywhere within the property lines, but it is desirable to minimize earthmoving and balance the volumes of cut and filled earth. Following are further constraints:

- Cut and filled slopes may not exceed a grade of 20 percent.
- Retaining walls may not be used.
- Indicate the level pad with solid lines and note its finish elevation.
- Ignore considerations of drainage.

184 Site Planning

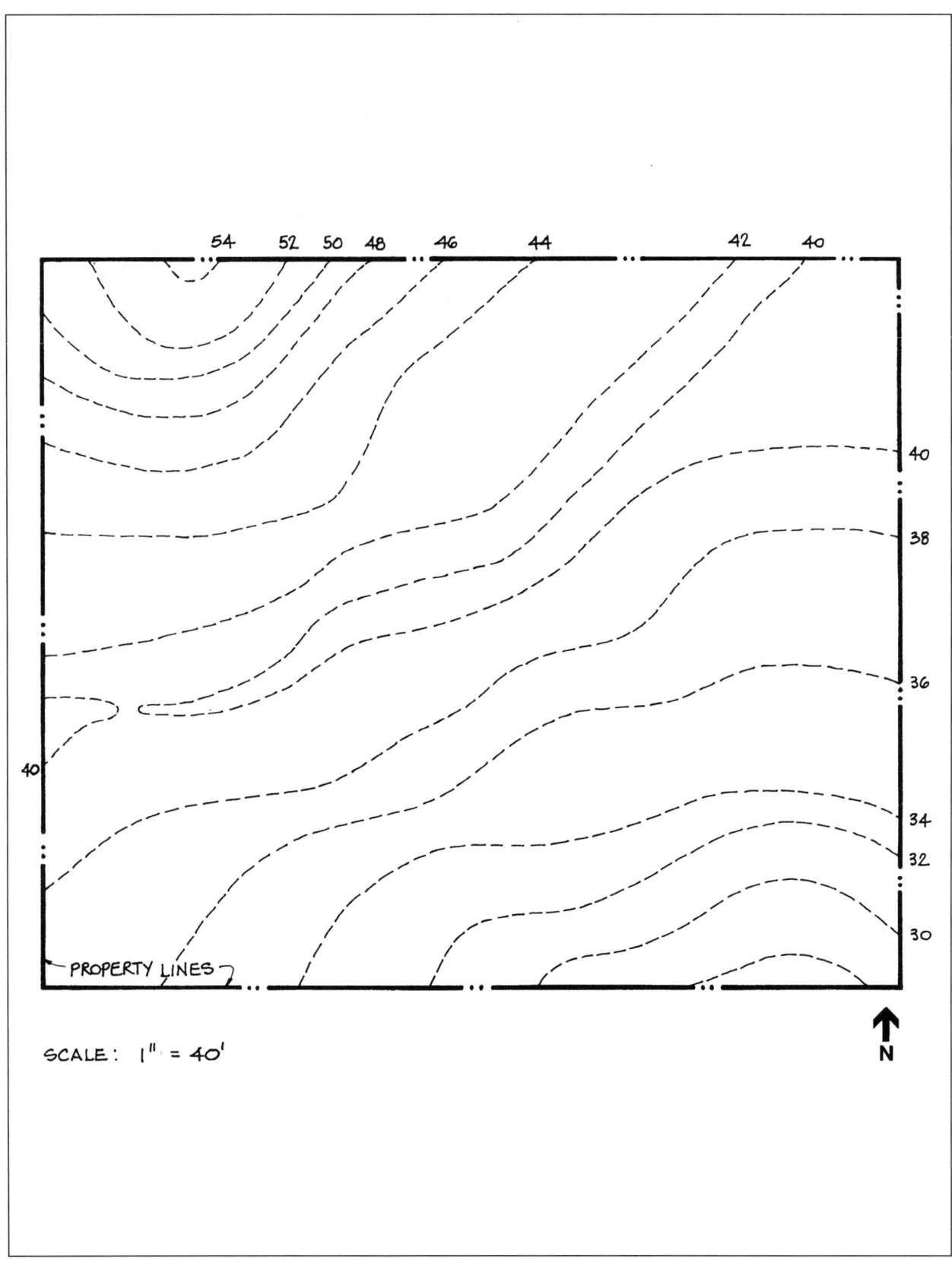

Problem 8: Discussion of Solution

Candidates experienced in reading topographic maps will recognize that this piece of property slopes rather unevenly from the northwest corner down to the southeast corner. The closely spaced contours at both these corners indicate a steeper grade than at the middle of the site, where the contours are more widely spaced. In fact, the two nearly parallel segments of contour 40 represent a strip of land that is relatively flat.

Locating the 80 × 120 foot level area is not a precise process. However, there are a couple of guidelines that should help direct one to a reasonable solution. First, the long sides of a rectangular pad should be approximately parallel to the existing contours. A plan shape that parallels the contours always requires less contour modification than a shape that runs perpendicular to the contours. That is why mountainous areas have winding roads rather than straight roads running directly over rugged terrain. The former follows the contours, that is, the natural shape of the land, while the latter opposes the contours and always requires far more grading.

Secondly, a rectangular pad should be placed on the flattest portion of the property. In addition, it should not be placed too close to any property line, so that all contours may be modified within the property borders. For example, if one side of the rectangle coincided with a property line, it would be impossible to modify any existing contour which intersected that line. For a clearer picture, see the diagram in the right column.

As previously mentioned, locating a pad is not an exact process, and there may be many locations which would be acceptable. However, by

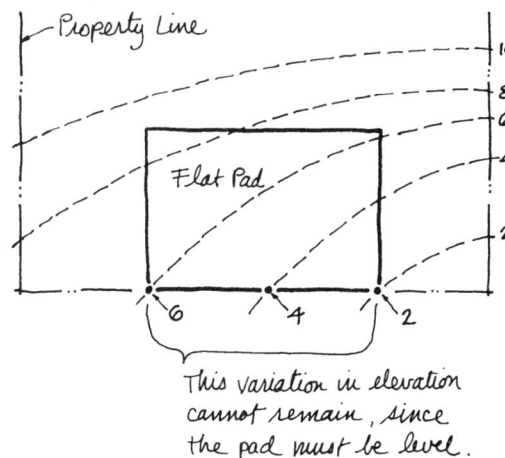

following the guidelines just discussed, we have located our flat pad as shown.

By drawing crossed diagonals within the rectangle, we determine that the center of this shape falls at elevation 39. We therefore set the finish grade of our level area at that elevation. Using crossed diagonals is a simple way to locate a midpoint, and on a regular slope, this will produce a finish pad elevation with balanced quantities of cut and fill. It does not work out quite as perfectly on an irregularly sloping grade (such as this one) but it should be good enough for our purpose.

With the rectangular pad located and its elevation established, we can now begin revising contours. The first contour higher than our pad is contour 40, and this revised line wraps around the high side of the pad, which indicates that we are cutting earth. Similarly, on the low side we take contour 38, the first contour below the pad elevation, and we wrap it around the low side, from the existing contour 38 on one side to the other. Revised contours along the low side indicate filled earth.

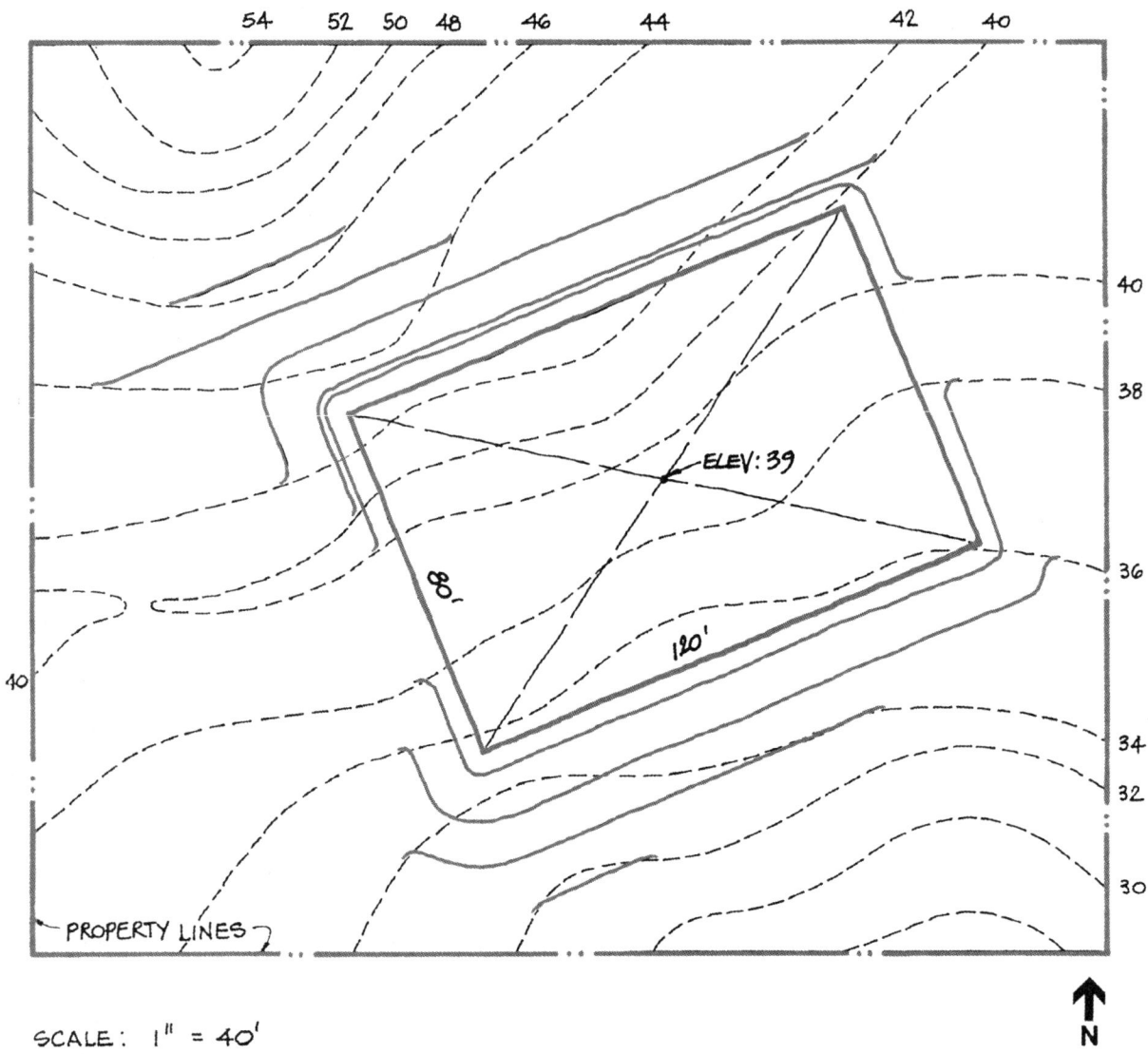

SCALE: 1" = 40'

The only question one may have about these revised contour lines is where they are drawn, relative to the edge of the flat pad. The problem statement specifies that modified slopes may not exceed a grade of 20 percent. Therefore, a contour that is either one foot higher or lower than the level pad must be placed a horizontal distance of 5 feet from the pad edge. Remember, $G = V/H$, where G is the gradient and V and H are the vertical and horizontal components. Thus, $H = V/G$ or $1 \div .20 = 5'$. Incidentally, these revised contours are drawn parallel to the pad edges, indicating a constant grade along the full length of the pad.

The next step is to revise the remaining contours on both the high and low sides. These are handled in the same way as previously described, with the following exception: Since subsequent contour intervals are 2 feet, the distance separating these contours must be 10 feet, in order to maintain the maximum allowable slope of 20 percent. All modified contour lines are drawn parallel to the sides of the level pad, and each is reconnected to its existing counterpart where the two lines intersect.

Contours are modified around the level pad until the last modified contour is at least 10 feet away from the next existing contour. At that point (in our example, between contours 46 and 48 and contours 30 and 32), the natural grade is already 20 percent or more, and hence, there is no further need to change the grade. In section, this will be where the cut or fill line meets natural grade.

One more aspect of this problem remains to be discussed, and that involves the troublesome matter of contour 40. Candidates are probably aware that this contour describes a relatively flat strip of land that runs through the center of the property. As such, it is composed of two separate legs that connect at the west side and diverge as they move eastward. Both segments of contour 40 must be considered, and one should keep in mind that contours may not cross, nor can they form a Y shape. Therefore, we have indicated one possible solution that employs two adjacent contours at the same elevation. This indicates a flat area between the two lines, whose constant elevation is 40. There may be other ways to handle this kind of problem, but this way is direct and reasonable.

Shown below is a section cut through the midpoint of our finished rectangular area. Although this was not a requirement of the problem, we provide it so that candidates may better visualize what is happening. Remember, if you are ever in doubt about contour modification, it is a relatively simple matter to cut a section and see the situation from another point of view.

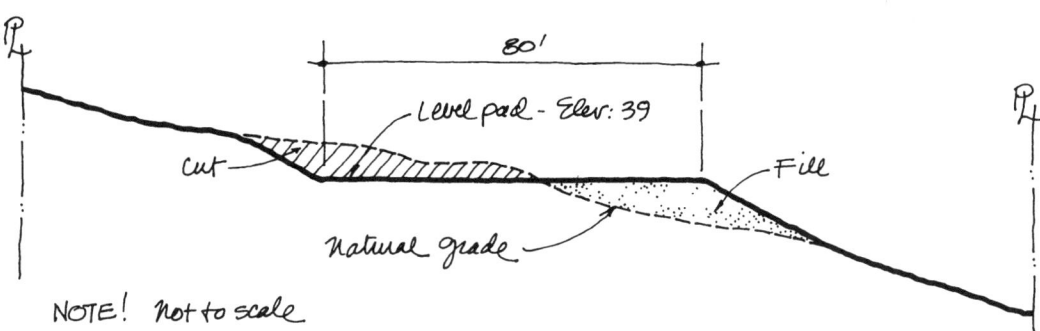

PROBLEM 9 PARKING

On the level corner site shown, you are asked to provide parking for 11 cars to serve the existing building.

Assignment:

Provide the following:

- 10 car spaces 9' × 18', except that
- End spaces shall be 12' × 18'
- One handicapped space 13' × 18'

Traffic shall be one way, with two 12-foot curb cuts for one entry and one exit driveway. Show traffic flow with arrows, number the parking stalls, and indicate all key dimensions. The setback areas shall be used for landscaping.

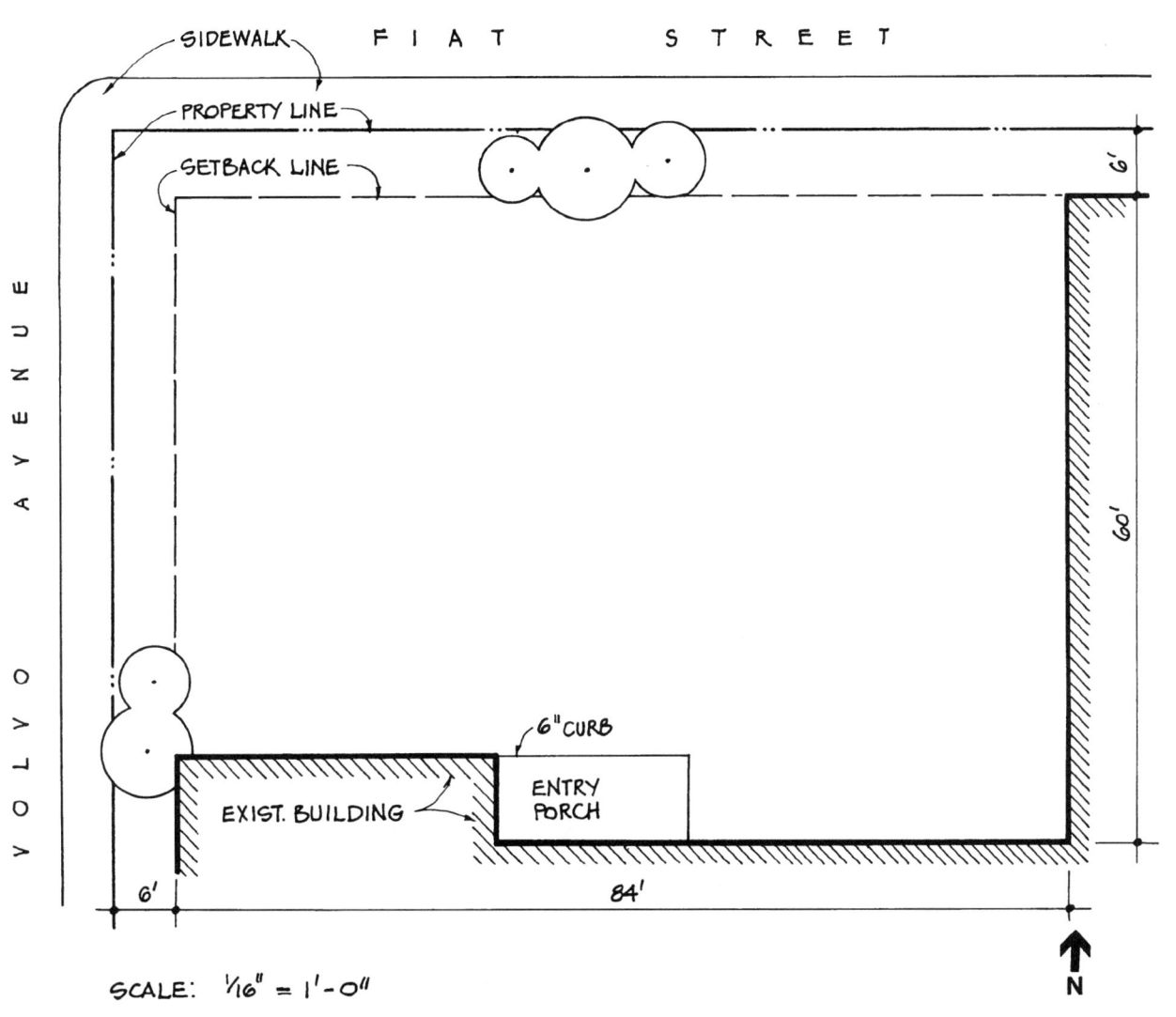

Problem 9: Discussion of Solution

For the past several years, at least one vignette problem on every test has dealt with parking. It should be clear to all candidates, therefore, that the examiners consider parking design to be very important.

Parking problems involve more than just the layout of a specific number of car spaces. Equally important is the circulation involved, and that means considerations of pedestrian conflicts, driver convenience, and, above all, safety.

In every past test, parking problems have involved 90-degree parking. Although a future problem could include angled parking, the odds seem to be against it. Therefore, when a problem (such as this one) specifies a certain size parking space, candidates should assume that the layout will involve 90-degree perpendicular parking. Since most problems include the sizes of parking spaces, there is only one other key dimension that one must remember. And that dimension is the required backup space of 24 feet, which is also the minimum aisle width between parallel rows of parked cars.

It is logical to begin any parking layout by placing car spaces along the perimeter of a site, because the longest dimensions of a site will accommodate the greatest number of vehicles. Another useful general rule is that driveways should be located as far as possible from street intersections. The reasons for this should be evident. First, the four-way traffic at an intersection often obstructs circulation to and from a corner driveway. Secondly, a car turning the corner (in this case, turning right onto Fiat Street from Volvo Avenue) must be aware not only of oncoming traffic, but also of cars entering or exiting the parking area. Finally, driveways near street corners are dangerous because, as cars slow to enter the lot, they force the traffic behind them to slow, which may block traffic on side streets. Cars exiting a lot may cause a similar interruption of cross traffic.

With all of the foregoing in mind, we may begin this simple parking layout. First, because two curb cuts are required, we assume that there will be one driveway at each street, and both driveways will be as far from the corner as possible. Two widely spaced driveways are always preferred, because that arrangement results in the most efficient vehicular circulation. If you placed the entry and exit driveways next to one another, with one 24-foot curb cut, you would have created a parking dead end, convoluted auto circulation, and the possibility of more dented fenders.

Secondly, we can see that the 60-foot lot width will neatly accommodate two perpendicular rows of cars plus a minimum aisle and backup space ($2 \times 18 + 24 = 60$). Therefore, we start by measuring a 12-foot driveway width from the building wall at the east (along Fiat Street) and then proceed to lay out as many car spaces as will fit along the northern setback line.

The first space (Space 1) is 12 feet wide (end space), and the remaining spaces (Spaces 2 to 6) are 9 feet wide each. The most westerly space (Space 7) is another 12-foot-wide end space. On the opposite side of the lot, we begin our layout with a 12-foot-wide end space (Space 8) at the southeast corner. Next, we place a normal 9-foot-wide space, and then our handicapped space (Space 10), which is located closest to the building entrance. So far, this has all worked out exceptionally easily. However, there is a catch: We are short one required parking space.

We know that a perpendicular parking space opposite Spaces 6 and 7 will not fit, because

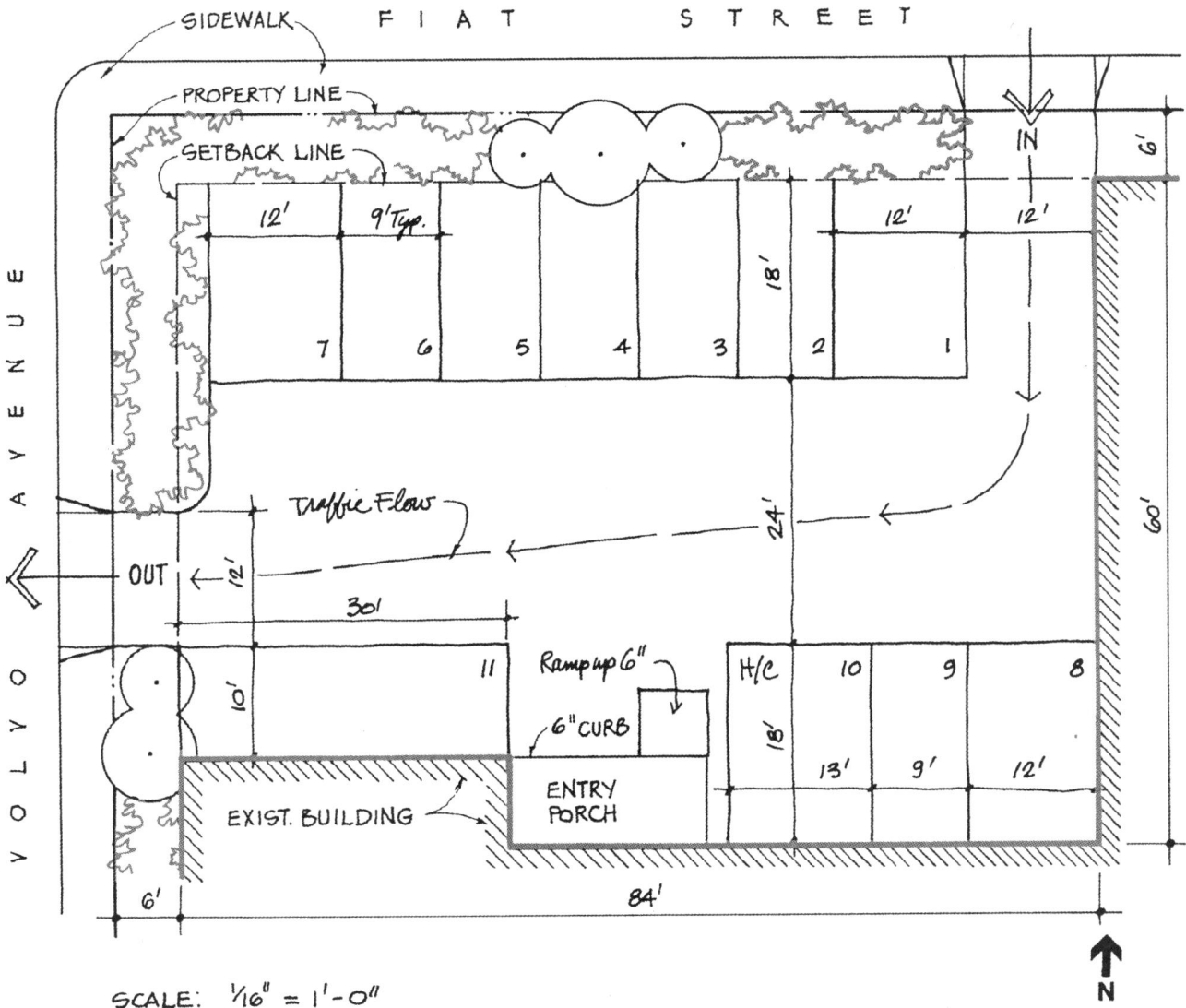

there is only about 34 feet behind the northerly row of cars, and we need at least 24 feet for an aisle plus another 18 feet for a car. At this point candidates may be tempted to abandon their solution and try another arrangement. Perhaps perpendicular spaces against the east building wall would work out better, and such an arrangement is shown on page 200.

As one can see, this solution creates even more problems and results in fewer total spaces.

However, there is another way to accommodate the missing car space. Although it is relatively unconventional, we can use one parallel parking space against the building wall at the southwest corner of the lot. We show Space 11 as a 10 × 30 foot area, because we have the space to do so. Actually, one could manage with a typical street-side parking space, which commonly measures 8 × 26 feet. Thus, with this unusual tactic, all eleven spaces are fitted on our restricted site.

We must still provide another driveway on Volvo Avenue, and this is placed as far from the corner as possible, at the southern edge of the traffic aisle. We now have to decide which driveway is the entry and which is the exit. Since we have no way of knowing if one or the other street is major or minor, our decision is based only on the benefit of permitting drivers who are unable to find a space to exit and re-enter the lot with a series of right turns. This entering on Fiat and exiting on Volvo arrangement also avoids the necessity of crossing traffic lanes, which would result from a reversed traffic flow, as illustrated in the next column.

A minor element of this problem is the 6-inch curb at the entry porch. For legal access, we must provide a ramp from the parking level to the level of the porch. At a maximum slope of 1:12, the 6-inch rise will require a ramp 6 feet long.

Since this ramp is no longer than six feet and it rises no more than six inches, it does not require handrails. In fact, those ramp dimensions are the maximum permitted without the requirement for a handrail. Finally, we must number the parking spaces and provide all the key dimensions. With

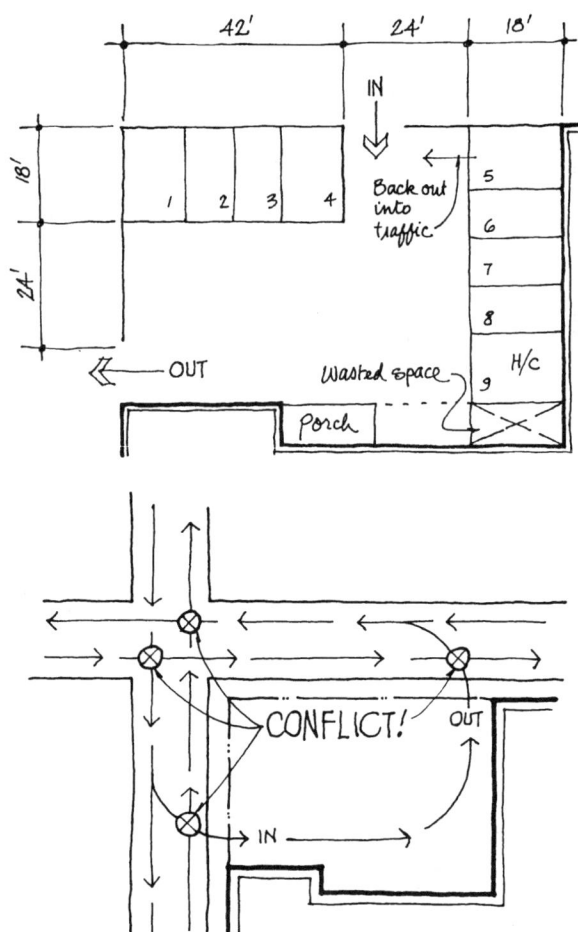

the indication of landscaping in the setback areas, our solution is complete.

PROBLEM 10 CONTOUR INTERPOLATION

Near the center of the site shown is an open paved area with a catch basin noted. At the east and west sides of the open paved area are 10-foot-wide pedestrian walks, while at the north is an area defined by dashed lines, within which fits an earth berm. The spot elevations represent finish elevations that must be maintained.

Assignment:

Draw continuous solid line contours 1 through 8 to produce a complete topographic drawing of the paved surfaces, unpaved surfaces, and earth berm. The location of contours must be interpolated from existing spot elevations, and the goal is to produce plane surfaces, which slope as evenly as possible. Finish walk grades may not exceed 5 percent, and other paved surfaces may not exceed a slope of 7 percent.

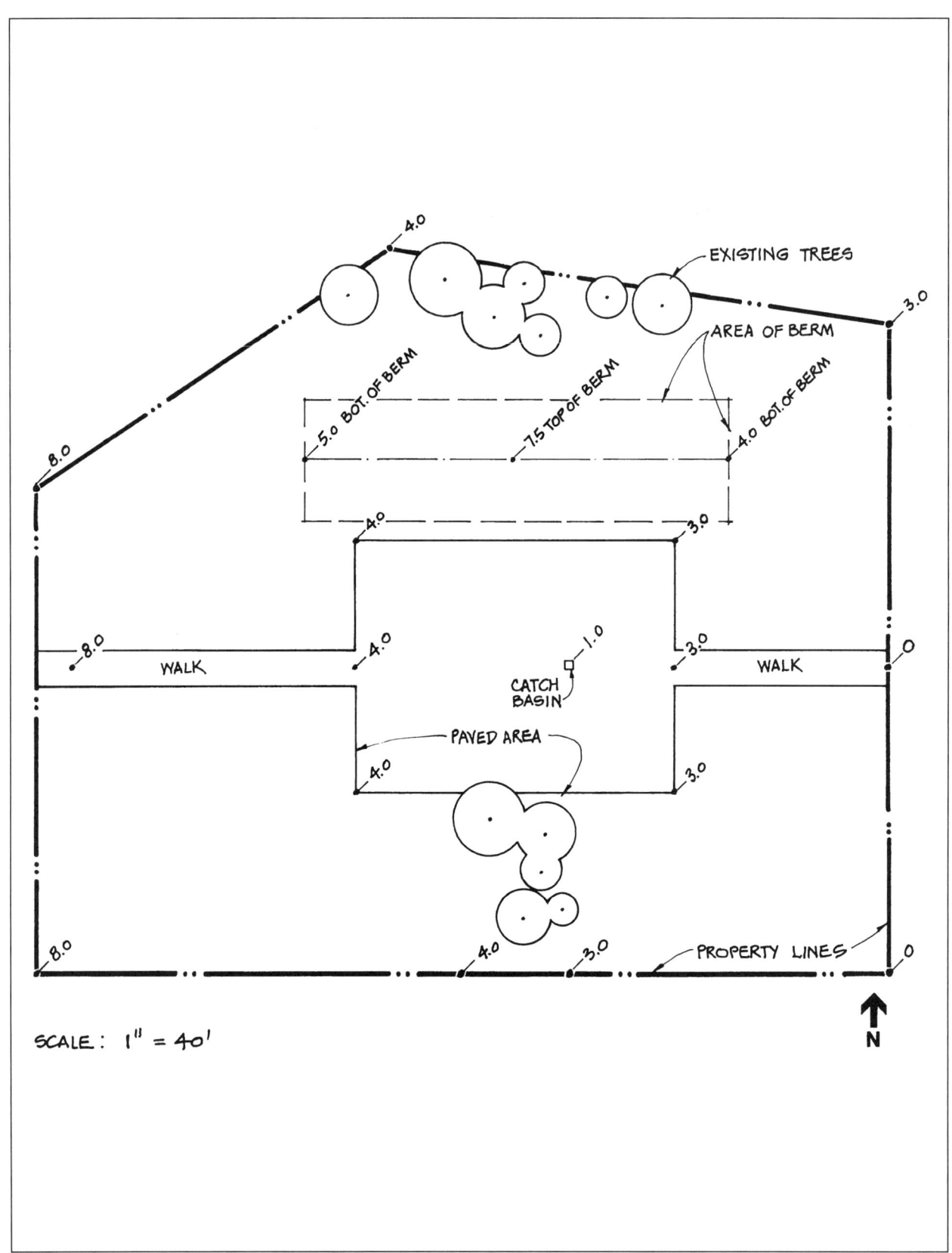

Problem 10: Discussion of Solution

In this exercise, we interpolate between known elevations to determine contours. For example, at the paved walk on the east side we begin with a known elevation of 0, along the east property line, and moving west, we rise to 3.0, where the walk meets the large paved area. We assume that perpendicular contours crossing this segment of walk will be evenly spaced, and since the length of walk scales 60 feet, the intervening contours will be 20 feet apart, as shown.

The slope of this walk, a three-foot rise in the 60-foot length, results in a 5 percent grade (3 ÷ 60 = 0.05), which is exactly the maximum grade specified in the description of the problem. The contours crossing the walk on the west side are determined in the same way, from spot grades 4.0 to 8.0, and thus, we have begun the process of solving this contour interpolation problem.

The next area to consider is the open paved area. We begin at the catch basin, whose spot elevation is noted as 1.0, and divide the distances evenly to spot grade 3.0 at the easterly walk and spot grade 4.0 at the westerly walk. These divisions determine the locations of evenly spaced contours that run north and south. However, contour 3, within the paved area, does not run to the north or south edge of the pavement. Rather, it must be connected to the northeast and southeast corners, which are also noted as 3.0.

The angle made by contour 3 must be such that the perpendicular distance from the catch basin to the contour line is 30 feet. This is because the slope of the paving may not exceed 7 percent (2 ÷ 30 = .067).

Contour 3, both north and south of the catch basin, should have the same angle in order to be symmetrical. Once this angle is established,

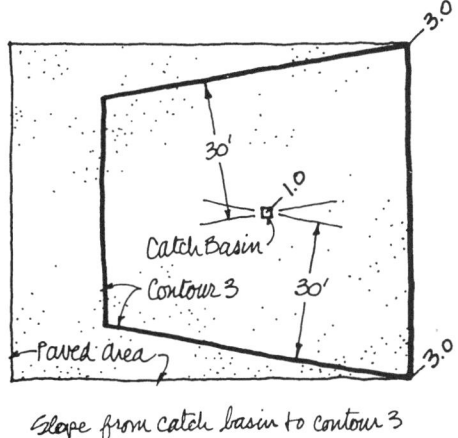

Slope from catch basin to contour 3 is $V/H = 2'/30' = .067 = \pm 7\%$

you must use the same angle for contour line 2, so that all contours within the paved area are parallel. This will assure a uniform slope on all sides and an even flow of water to the catch basin. Remember, evenly spaced contours always represent a uniform slope, and straight line contours that are evenly spaced always represent a plane surface.

The contours in the paved area east of the catch basin are about 15 feet apart, resulting in a grade of just under the prescribed 7 percent (1 ÷ 15 = .067). The contours in the paved area west of the catch basin are about 20 feet apart, and therefore, flatter than 7 percent. The distance between contours 2 and 3, both north and south of the catch basin, are also drawn 15 feet apart, and thus, they conform to the program requirements.

Since contour 3 within the paved area is completely enclosed, the other contour 3 that begins at the site's northeast corner cannot be connected to it. Remember, contours may never split in two. Instead, we run an adjacent contour 3 parallel to the east side of the paved area and then southwesterly to spot elevation 3 along the southern property line. The area between these

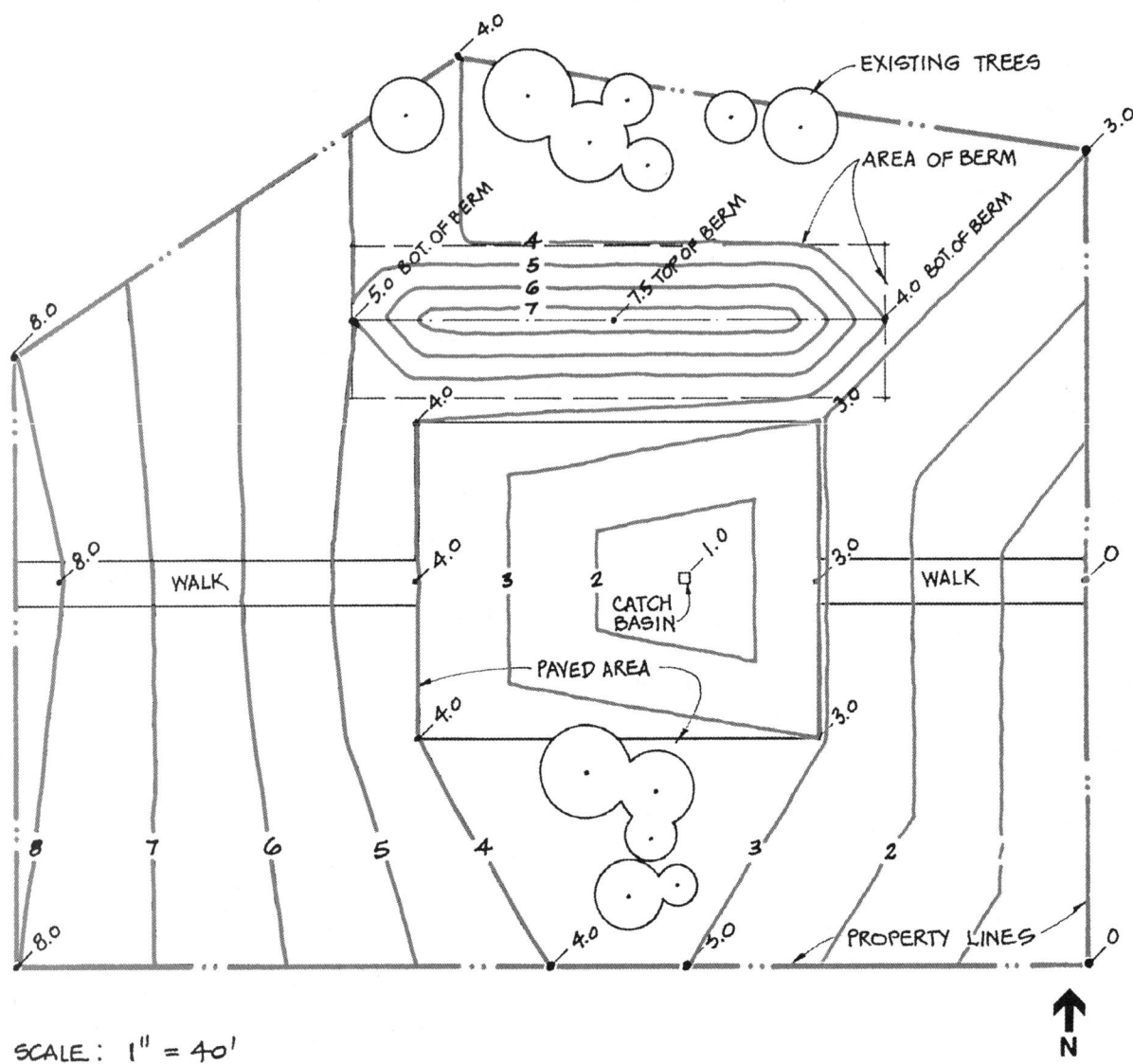

SCALE: 1" = 40'

closely spaced contour lines is virtually level, since they both represent an elevation of 3.

Next, we can establish the finish contour points along all the property lines. As before, we must interpolate these points using the spot grades already known. Thus, across the southerly property line we establish every contour point, which conveniently works out to be at equal 30-foot intervals (240 feet ÷ 8 spaces = 30 feet). The other sides are handled similarly, except for the lower half of the east line and the most northerly line, both of which have no intervening contour points.

At this point, we can begin to complete some of the contour lines, such as lines 6, 7, and 8 at the west, which are relatively straight. At the east side, we begin with spot elevation 3, at the northeast corner of the site, and as previously described, this contour 3 runs toward the open paved area. The angle established by contour 3 is duplicated for contours 1 and 2. Remember, our goal is to produce even slopes.

The only major piece of this puzzle that still remains is the earth berm. The berm area is defined by dashed lines, and in addition, we are given spot elevations of the top and both east and west ends. We develop an elongated diamond shape for our berm, because we prefer to avoid the intersection of contour 4 at the east side with the angled contour 3 that was just drawn to the northeast corner of the site. Those who elect to fit their berm completely within the dashed line area should end up with a configuration that resembles the following, which is equally acceptable:

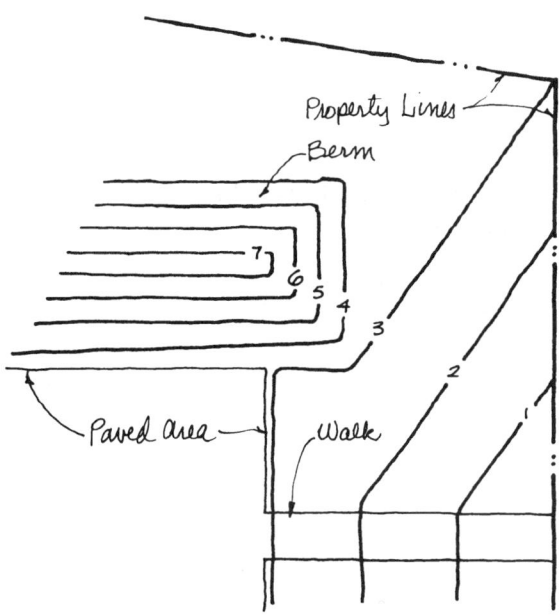

In any case, with the berm top noted as 7.5, our highest contour will be 7, because our contours have even one-foot values. Therefore, we divide the berm area into equally spaced contours that start near the top at 7 and fall away in all directions to 4 at the east and 5 at the west. Note that the easterly spot elevation of 4, at the bottom of the berm, connects to spot elevation 4 at the northwest corner of the paved area. This is the only practical way that contour 4 can be continuous.

Similarly, contour 5, at the west end of the berm, diverges; the upper end of the contour goes straight north, the middle portion of the contour travels around the berm, and the lower end of the contour goes south. There is an alternate solution to this situation, which has two separate 5 contours and appears as follows:

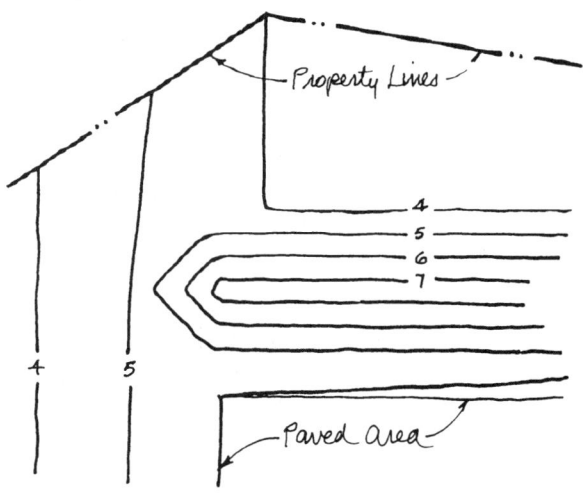

With the berm completed, we now connect every segment of every contour, so that each becomes a continuous line. In addition, we make each contour as regular and straight as possible so that all slopes will be relatively even, as required by the program. As one can see, the only practical way to solve a problem of this kind is one small part at a time. Since final contours are often uncertain until the very end, one may have little idea of how the overall solution should appear until the last line is drawn. During the process, however, one must try to anticipate the next step in order to avoid costly errors in the final solution.

PROBLEM 11 PARKING AND DRAINAGE

Given the existing site with the outline and finish spot grades for a new parking lot. At the south is an existing access road with drainage ditches on both sides, and to the west is located a new building for which the new parking is to be provided.

Assignment:

- Regrade the site to accommodate the new parking lot with finish spot grades for paving, as shown.
- Divert runoff water around the parking lot and towards the existing drainage ditch.
- Show new contours on the paved lot with a maximum finished grade of 5 percent.
- Provide 17 - 10' × 20' car spaces and 3 - 15' × 20' handicapped car spaces.
- Provide a 20-foot-wide driveway, with a maximum finished grade of 10 percent, from the road to the new parking lot.
- Provide a 5-foot-wide pedestrian path, with a maximum finished grade of 1:20, from the parking lot to the new building.

Lesson Ten: Exercise Problems **199**

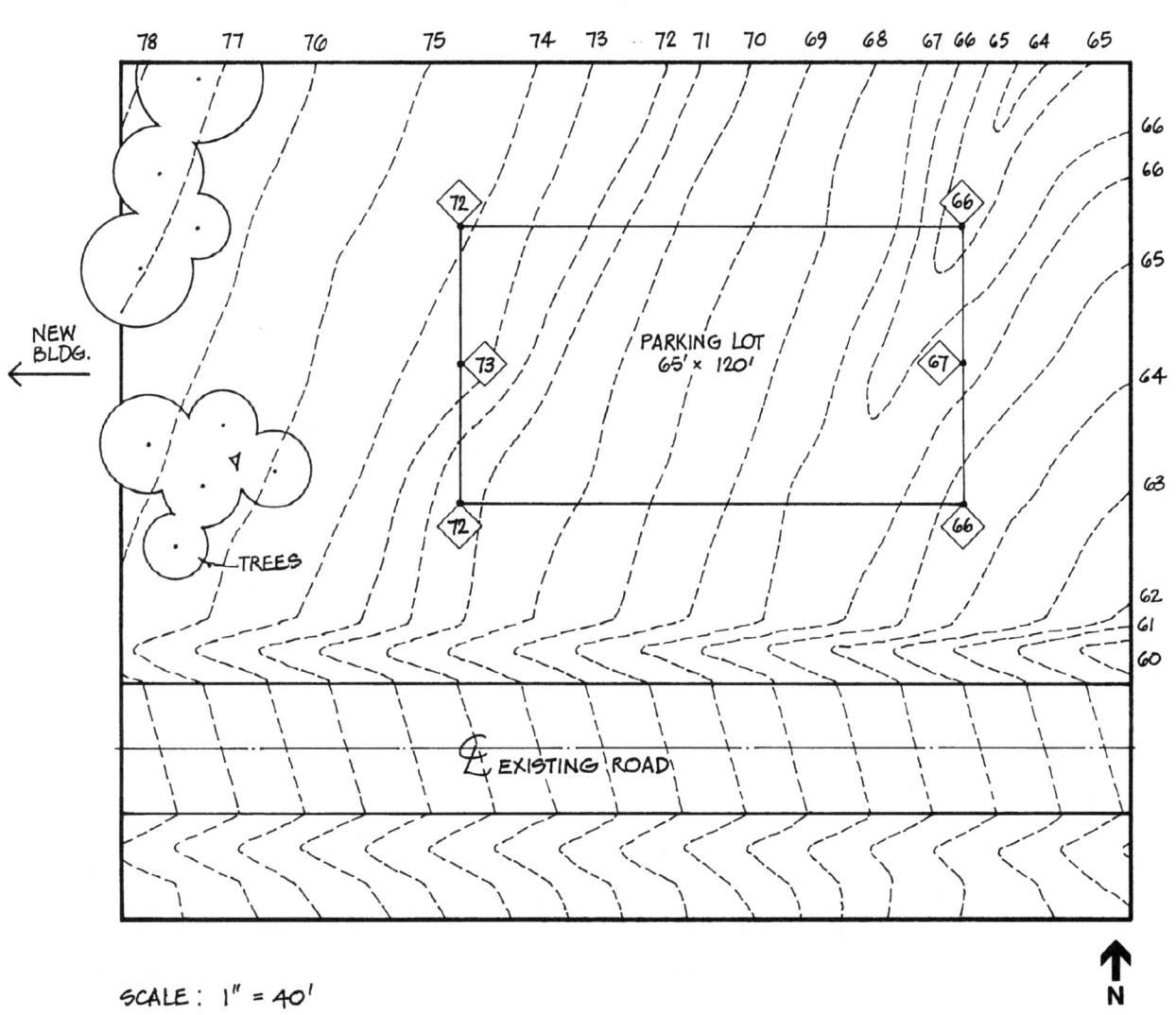

SCALE: 1" = 40'

Problem 11: Discussion of Solution

This is a complicated problem that requires a number of individual steps to complete. No one part is particularly difficult, but developing all the necessary patterns of parking and drainage is intricate and time-consuming. Candidates should probably allow at least 45 minutes to do all this work.

As in all graphic site problems, one should review the existing conditions to get a sense of what's going on here. We see a large land parcel that has a relatively steady slope downward from west to east. At the south, there is shown a 30-foot-wide paved road, and on both sides and parallel to this road are drainage ditches that carry water eastward. We also see the outline of a new parking lot, complete with elevations at the corners, as well as at the midpoints of the short sides.

Normally, it is best to solve the architectural problems first, so that every element is in place before contour modification begins. In this case, that means we must begin by arranging the 20 required parking spaces and adding the necessary driveway and pedestrian path. The placement of the driveway and path is determined primarily by the location of the new building. Since this building is off the page at the far west, our path should be a straight line connection westward from the parking lot. Thus, we center the pedestrian path on the west side of the parking lot and run it due west between the two groups of existing trees. With the path established at the west, it seems most logical to place our handicapped parking spaces on the west side as well, so that handicapped drivers will be as close as possible to the new building.

In locating the required driveway from the road to the parking lot, we have three choices. First, we could place the driveway at the far west side of the lot, but in that location it would displace at least one handicapped parking space, and it would also create a hazard for pedestrians approaching the path to the new building. The second choice is to locate the driveway in the middle of the south side of the lot, but that would result in traffic dead ends at the east and west ends of the parking lot. The final choice is to locate the driveway at the far east side of the lot, which is shown in our solution. At this location we avoid the pedestrian path, we eliminate the conflict with the handicapped spaces, and we create only one traffic dead end.

The actual layout of car spaces is always a trial-and-error exercise. In this case, where the lot width is 65 feet, two parallel rows of cars, 25 feet apart, fit very conveniently. There might be other suitable layouts, but this arrangement is direct, simple, and, most importantly, it works. Note that space 11 is 15 feet wide, first because the space was available, and second, because end spaces generally need additional width for maneuvering.

Although it was not previously mentioned, parking problems invariably comprise 90-degree parking. One should never consider angled parking unless it is specifically required. And in this case, where there is only one way in and out of the lot, angled parking would be highly impractical.

Our next problem is to regrade the site and parking lot in order to divert the runoff water toward the existing drainage ditch. We begin with the finished spot grades shown on the paved area. In analyzing the grades along the short sides of the lot, we discover that the paved lot is one foot higher at the center than at the ends of each side. This should immediately indicate to candidates that the paved lot will have a central crown, and

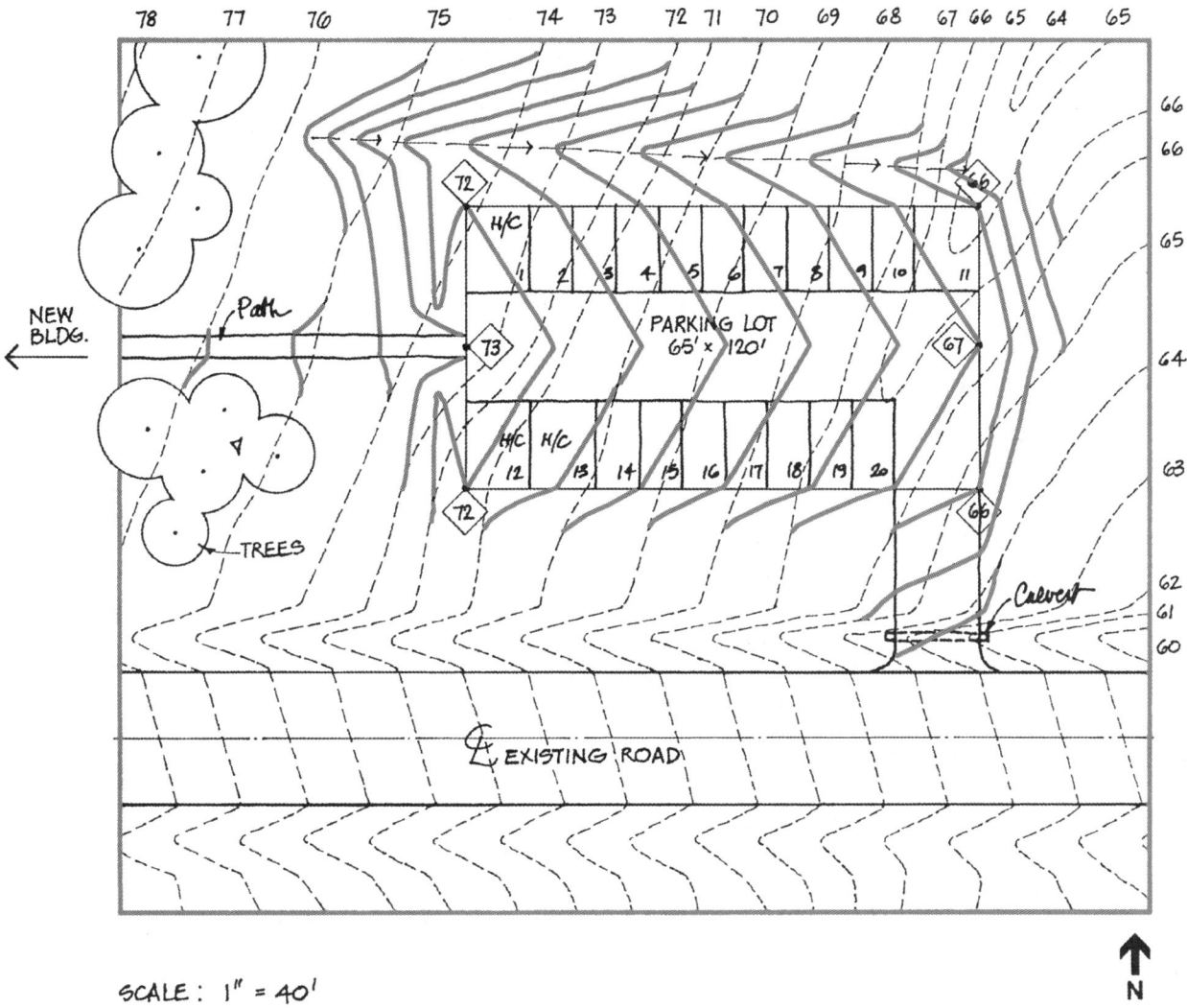

SCALE: 1" = 40'

the surface water will flow both northeast and southeast towards the long sides.

We begin with spot elevation 67, at the center of the east side, and divide the long centerline of the lot into six equal spaces to reach spot elevation 73 at the center of the west side. These points represent the tops of the continuous paved crown, and each of these is conveniently spaced 20 feet apart, resulting in the required finished grade of 5 percent (1:20). The tops of the crowns are then connected to their corresponding grades at the north and south borders of the lot, and this creates the chevron-like pattern one sees here.

Incidentally, if one were to calculate the grade based on the perpendicular distance between contours, the actual finish grade would slightly exceed 5 percent, but we feel that would probably be acceptable in this case. The following sketch illustrates this situation.

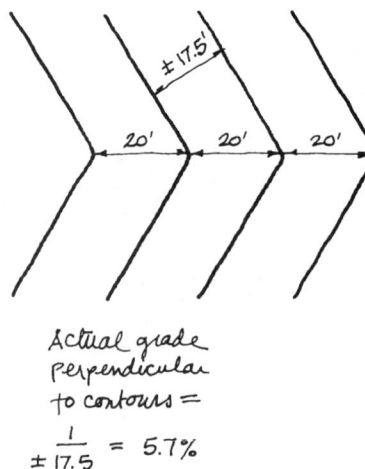

Actual grade perpendicular to contours = $\dfrac{1}{\pm 17.5} = 5.7\%$

We next regrade the site outside the paved area to conduct the runoff water around the parking lot. To accomplish this, we must develop a drainage swale at the north that will run from west to east. If we begin with spot grade 72 at the northwest corner of the lot, and pull that contour southward toward the new path, we will create a shallow drainage ditch that will prevent water from flowing onto the parking lot from the higher elevations. We direct this same contour northward to form another loop and then reconnect it to the existing 72 contour near the north property line.

The next contour to be revised is 66 at the northeast corner of the lot. We form a loop north of the parking lot and then bring the contour southward around the lot, where it intersects the spot grade at the southeast corner of the lot. Once the 67 and 72 contours have been drawn, the distance between the two (representing the flow line) may be divided into roughly equal spaces to create intervening swale loops. This equal spacing insures that surface water will flow uniformly across the north portion of the site. At the west end, we must add additional swale loops at every affected contour, in this case, all the way through 75.

At the south side of the site, the parking lot contours must be reconnected to their corresponding lines, but it is not necessary to create another swale. Because of the direction of the revised contours, surface water will flow naturally toward the existing drainage ditch. The one exception, of course, is contour 72 at the south, which is drawn as a mirror image of the 72 contour along the parking lot north of the path.

The final step in this long problem is to indicate the required grades at both the driveway and the pedestrian walk. The contours cutting across the driveway are placed 10 feet apart (10 percent grade), and they are angled so that water will run in the southeast direction into the drainage ditch. The contours cutting across the path are placed 20 feet apart (1:20 or 5 percent grade) and then reconnected to their corresponding lines. Finally, we must indicate a culvert running beneath the driveway, to show that free flow of water in the drainage ditch has been recognized and resolved.

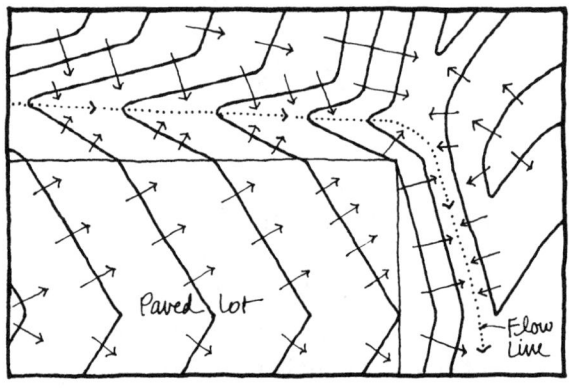

Verifying the Surface Flow

If you have followed this explanation and solved the problem as described, you will have little trouble going through a final check of your work. This is done by tracing every modified contour to be certain that each begins and ends at its corresponding line within the confines of the site. For example, the easterly 65 contour, second from the right, begins at the north with a small loop, wraps around the east side of the parking lot in a ridge-like configuration, crosses the driveway at an angle, and finally is reconnected to the existing 65 contour near the upper part of the drainage ditch.

Another useful test is to trace the direction of flow to be certain that the surface water is running exactly where it should. Candidates must remember that water always flows perpendicular to the contours. Therefore, if you draw small perpendicular arrows at each contour, as shown on the previous page, you will reveal any flow problems that might be present.

PROBLEM 12 GRADING

Given the topographic plan which includes a road layout and an existing drain running beneath the pavement. Also shown is the required finished road profile.

Assignment:

Draw revised contours with a solid line, incorporating the required road profile, and indicate the appropriate grading.

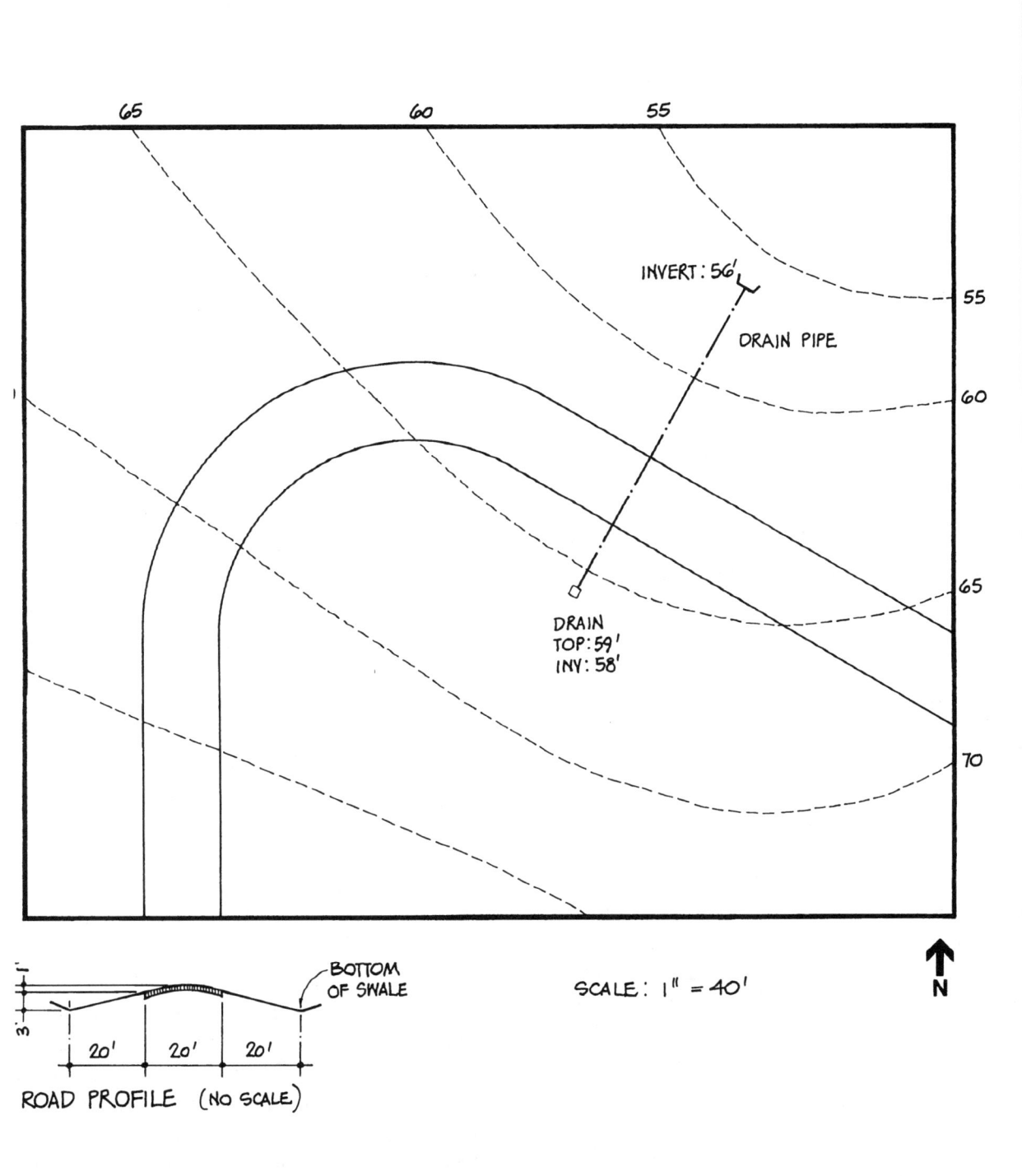

Problem 12: Discussion of Solution

Although it is relatively simple in concept, this problem requires a number of careful grading considerations. To begin with, we have a 20-foot-wide road that begins at the southwest corner of the site and runs downhill, as it makes a turn greater than 90 degrees. At the end of the turn somewhere near the drain line, it reaches a low point and then begins to run uphill, as it continues off the page towards the southeast.

One should also note that the required road profile indicates a one-foot-high central crown and a drainage swale on each side of the road. The flow lines of these drainage swales are located 20 feet from the pavement edges, and the depth of the swales is noted as three feet. The only other data influencing our solution are the drain elevations, which are indicated as 59 feet at the top, or inlet, 58 feet at the invert, or lowest inside surface of the pipe, and 56 feet at the invert elevation at the end of the drain, north of the road.

Before embarking on a solution, candidates should understand the entirety of this problem. The contours crossing the road must be modified, a crown in the road must be indicated, and drainage swales must be designed to carry the water along specific flow lines on both sides of the road. At the south, the surface water should be directed to flow towards the existing drain inlet, and at the north, the surface water should flow in the direction of the drain pipe termination, noted on the plan as invert: 56'.

One should begin the solution to this problem by revising the contours on the paved road. It is generally advisable to deal with the architectural, or man-made elements, first, since these features usually have the least flexibility. In this case, the road width is established, the height of the crown is fixed, and we assume that the gradient of the road must be as uniform as possible. Thus, we start by modifying the contours at both ends of the road.

We first determine where the 75 contour crosses the pavement, and we bisect that contour with a line drawn perpendicular to the sides of the road. Next, we do the same with the 65 contour at both locations where it crosses the road. In this case, the 65 perpendiculars are nearly equidistant from the line of the drain pipe, which is desirable, because we have assumed that the pavement above the drain pipe coincides with the road's low point. If these two perpendiculars were not equidistant from the drain pipe, we would have to modify these distances to make them more equal. The next road contour to be modified is 70, and the perpendicular crossing the road here should be placed equidistant from the 65 and 75 perpendiculars, which are immediately north and south of 70. Remember, equal distances between contours produce a uniform road gradient.

Since we have revised every contour that crosses the road, we may now proceed by indicating the crown in the road. Candidates might recall that a crown in the road is the same topographic configuration as a ridge, and a ridge is represented by contours pointing downhill. Therefore, a convex curved ridge is shown in plan as the arc of a circle that curves in the downhill direction. But what is the extent of that curve? The top of the curve is determined by the established dimensions, which in this case consist of a one-foot crown and a five-foot contour interval. In other words, if the distance between contours represents five feet vertically, the representation of a one-foot-high crown will appear in plan as a curve whose apex is one-fifth the distance between contours. So if you measure one-fifth the distance between the

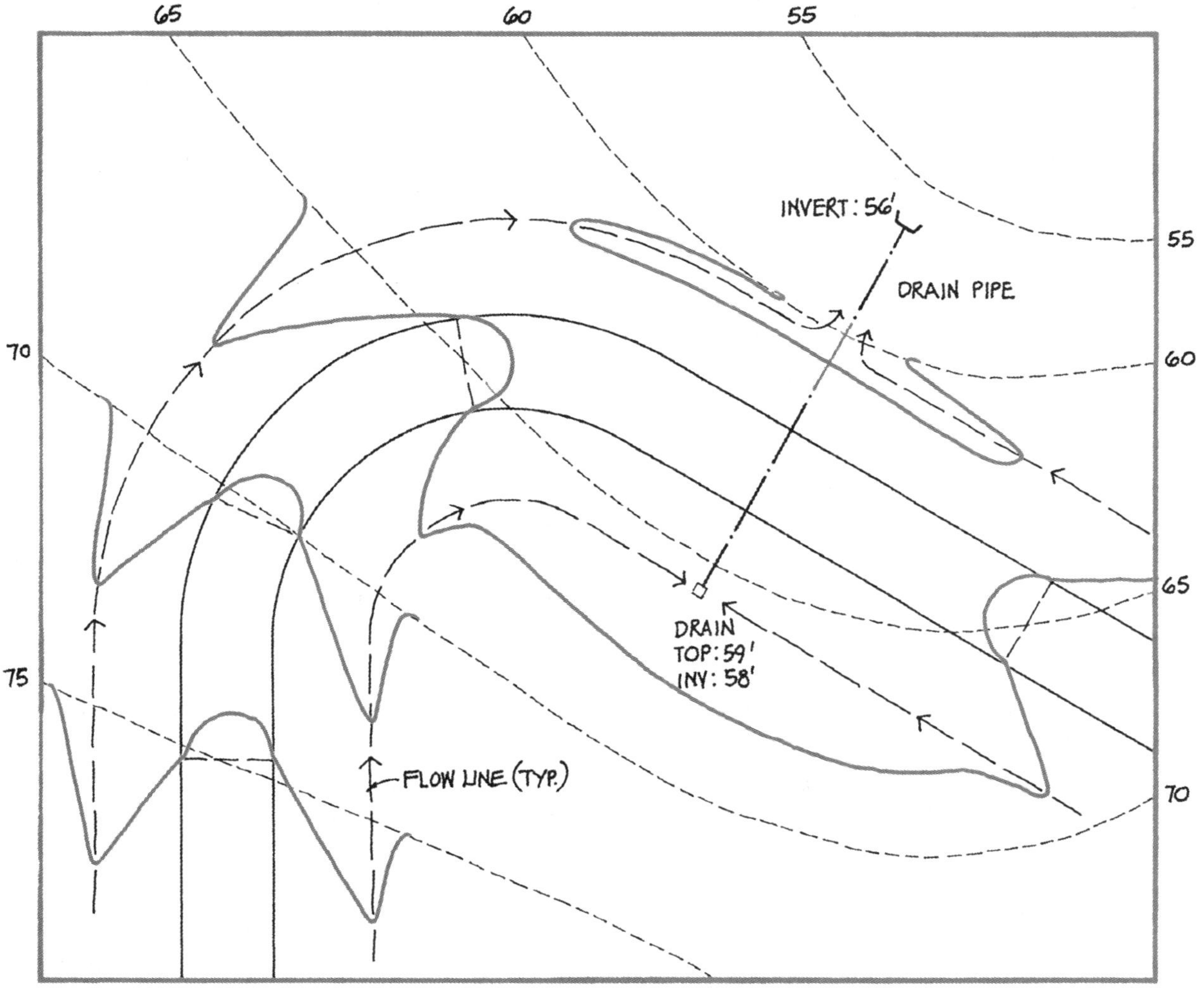

Lesson Ten: Exercise Problems   207

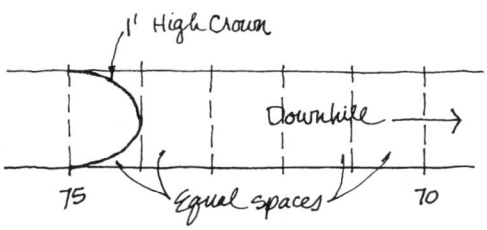

Plan of Road

road perpendiculars just established, you will find the top of a curve that must be drawn at each perpendicular. Note the diagram on the previous page.

One should also note that the crowns of contours 75, 70, and 65 at the west all point up and/or to the right, while the crown of contour 65 at the east points towards the left. This signifies that the slope of the road changes direction around the drain pipe, and therefore, the crown points downhill, as it should, in each case. Incidentally, what if you ignored the curves and simply used the straight line perpendiculars? In that event, you would have portrayed a level road section, rather than curved, and your score would have been correspondingly marked down. On the other hand, what if you simply guessed wrong about the curves, and they were drawn pointing uphill? In that event, you would have portrayed a saucer-like road configuration (contours pointing uphill represent a valley), and again, that aspect of your solution would have been incorrect and penalized accordingly. Note the following diagram:

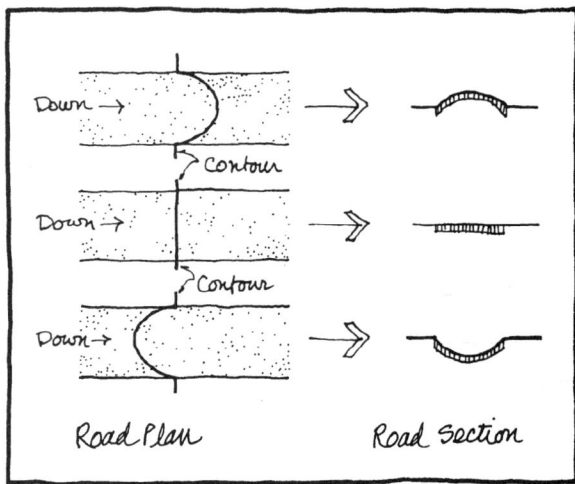

Our next step is to create flow lines for the swales indicated in the road profile. These are drawn in plan as dotted lines parallel to each side of the road and 20 feet from the pavement edges. To create the required swales, we may begin at either side of the road to form the loops representing the tops of the revised contours. However, these loops should be positioned approximately 3/5 the distance between the five-foot contours, in order to correctly represent the swale's three-foot depth. The concept is exactly the same as in locating the road crowns, except the contours in this case must point uphill. We point the contours uphill because we are creating a valley through which the surface water will flow.

Next, the swale loops are connected to their corresponding road crowns on one side, and on the other side they are reconnected to their corresponding contour lines. The one exception is contour 65, where the revised contour is pulled south of the drain so that water will flow downhill (northward) towards the drain. Finally, we revise contour 60, north of the road, to create two swales that guide the surface water along the flow line and in the direction of the drain pipe termination.

For candidates who wish to understand what happens to the contours around the drain inlet, we

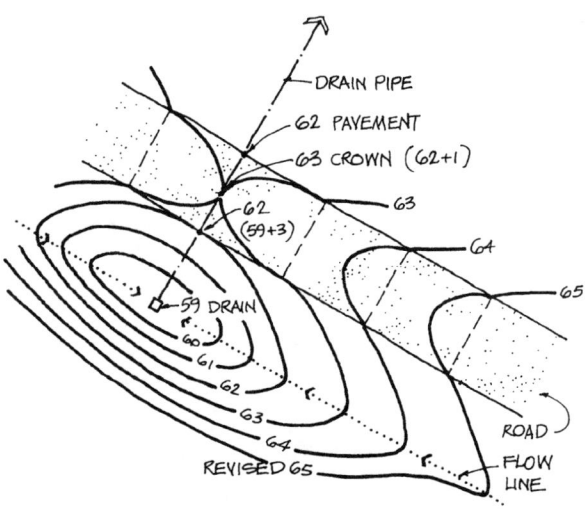

provide the sketch found at the bottom of the previous page, which shows the approximate position of one-foot contours.

A section cut through the drain pipe would show that the drain inlet is situated at the lowest point in the depression, indicated by the concentric contours. Although this information is not actually a part of this problem, candidates should understand how topography is revised to achieve certain desired results. The point is, contour revisions are not at all arbitrary. The slightest alteration of a contour line can result in a change in surface water flow, as well as moving of a considerable quantity of earth.

PROBLEM 13 SETBACK AND HEIGHT LIMITS

You have been retained to design a two-story building on the narrow plot of land shown. A typical story will be 12 feet in height (including construction depth), and your design will enclose the maximum building volume permitted. Setbacks are 20 feet at the front and 30 feet at the rear. However, any portion of the new structure which is less than 10 feet above the street must observe a setback of 30 feet. The maximum allowable building height is 12 feet above the highest elevation of the property, and no construction is permitted below existing grade.

Assignment:

On the section/graph provided, indicate the following:

- A cross section through the center of the site showing the existing grade.
- The maximum two-story structure permitted. Show by outline and hatching.
- Setback dimensions at front and rear for all levels of the structure.

Lesson Ten: Exercise Problems

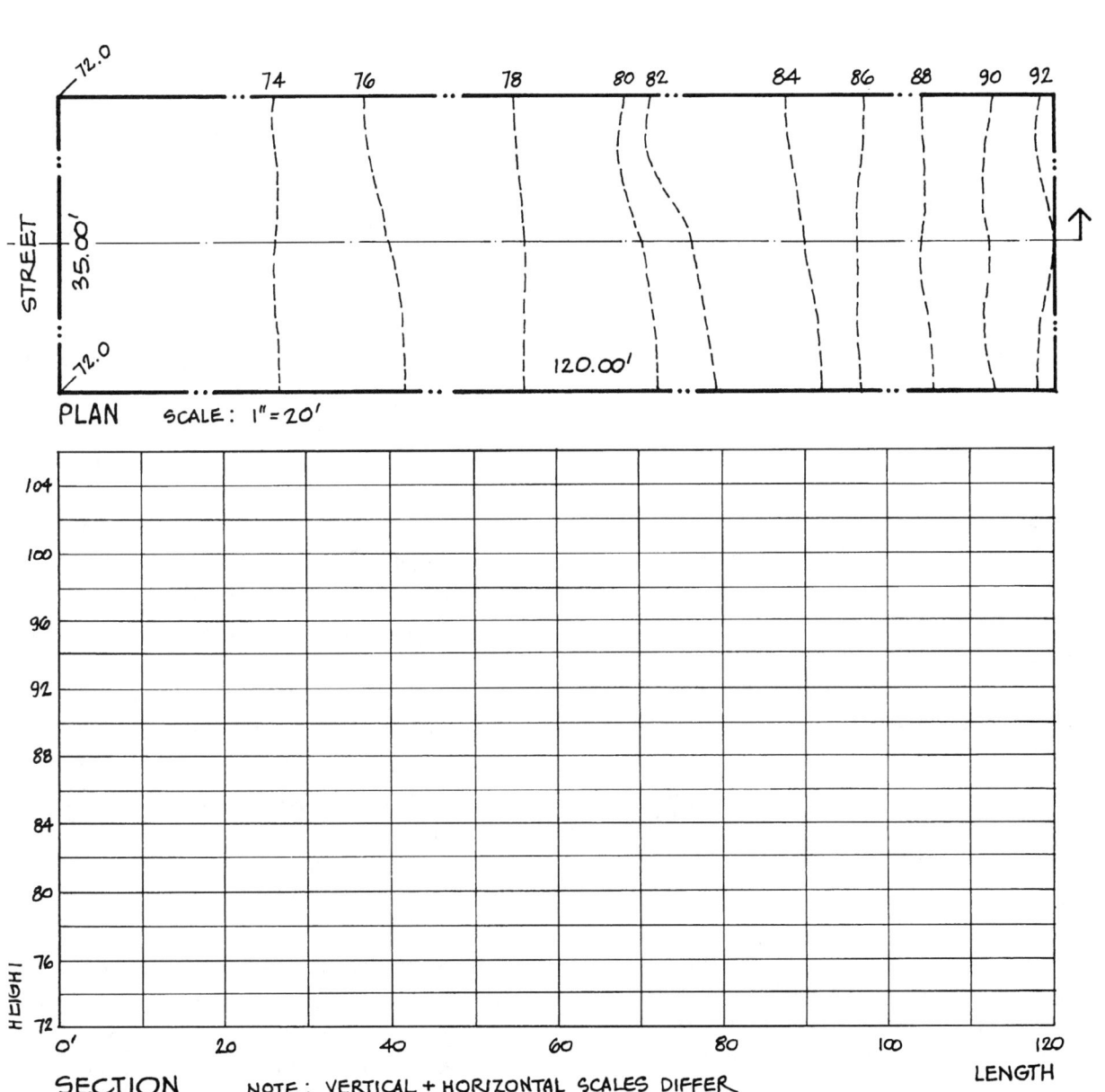

PLAN SCALE: 1" = 20'

SECTION NOTE: VERTICAL + HORIZONTAL SCALES DIFFER

Problem 13: Discussion of Solution

This is actually a development restriction type of problem in a slightly modified form. Instead of the usual site plan with restrictions that limit the buildable ground floor area, this problem has a combination of horizontal and vertical restrictions that limit the volume that may be built. Additionally, while maximum buildable area is calculated on the plan, maximum buildable volume is calculated on the section. Otherwise, the two problem types are quite similar. As in all development restriction problems, a candidate must take it one step at a time, interpret each written restriction, portray it graphically, and determine the maximum building volume that will result from all the applicable restrictions.

Before determining the shape of the maximum building volume, we must first draw a cross section through the center of the site. This section is drawn directly on the section/graph provided. Please note that the vertical and horizontal scales on this graph are different; that is, one inch of height represents 10 feet, while one inch of length represents 20 feet. Therefore, the shape of the resulting volume will be distorted and not to scale; it will be merely an abstract representation of the final shape.

One draws the required section by projecting points from the plan above at each intersection of an existing contour with the longitudinal center line. Thus, at every intersection of a contour with the center line one makes a dot, and a line from that dot is projected downward until it intersects its corresponding line below. For example, the intersection of contour 74 and the longitudinal center line is extended downward until it meets the horizontal 74 line on the graph on the previous page. This occurs about 26 feet to the right of the existing street, and at that point a dot is marked.

Each contour is marked similarly, projected downward, and noted on its corresponding horizontal line on the graph. When all of the points are projected by means of dots onto the graph on the following page, the dots are connected with a continuous, free-flowing profile line, which represents a section cut through the center of the property. Once we know the profile of the property, we can begin the second phase of the problem—applying the necessary restrictions.

The first restrictions are front and rear setback lines, which are marked along the 20-foot vertical line and 90-foot vertical line of the section/graph. It should be understood that the 90-foot vertical line represents the required 30-foot rear setback (120' − 90' = 30'). We are also told that any portion of the structure which is less than 10 feet above the street must have a 30-foot front setback. Since the street is at elevation 72, 10 feet above the street is at elevation 82. Therefore, along the horizontal 82 line on the graph we move to the right 30 feet from the street. We must remember that no structure may encroach within the area below elevation 82 and left of the 30-foot vertical line. In addition, elevation 82 becomes the floor level closest to the street.

Our next restriction is the overall building height, which was described as 12 feet above the highest elevation on the property. Since the highest elevation is about 92, our building may not project above elevation 104 (92' + 12' = 104'). Again we draw a horizontal limit line at elevation 104. In all these steps, we are establishing maximum limit lines beyond which our new building may not encroach.

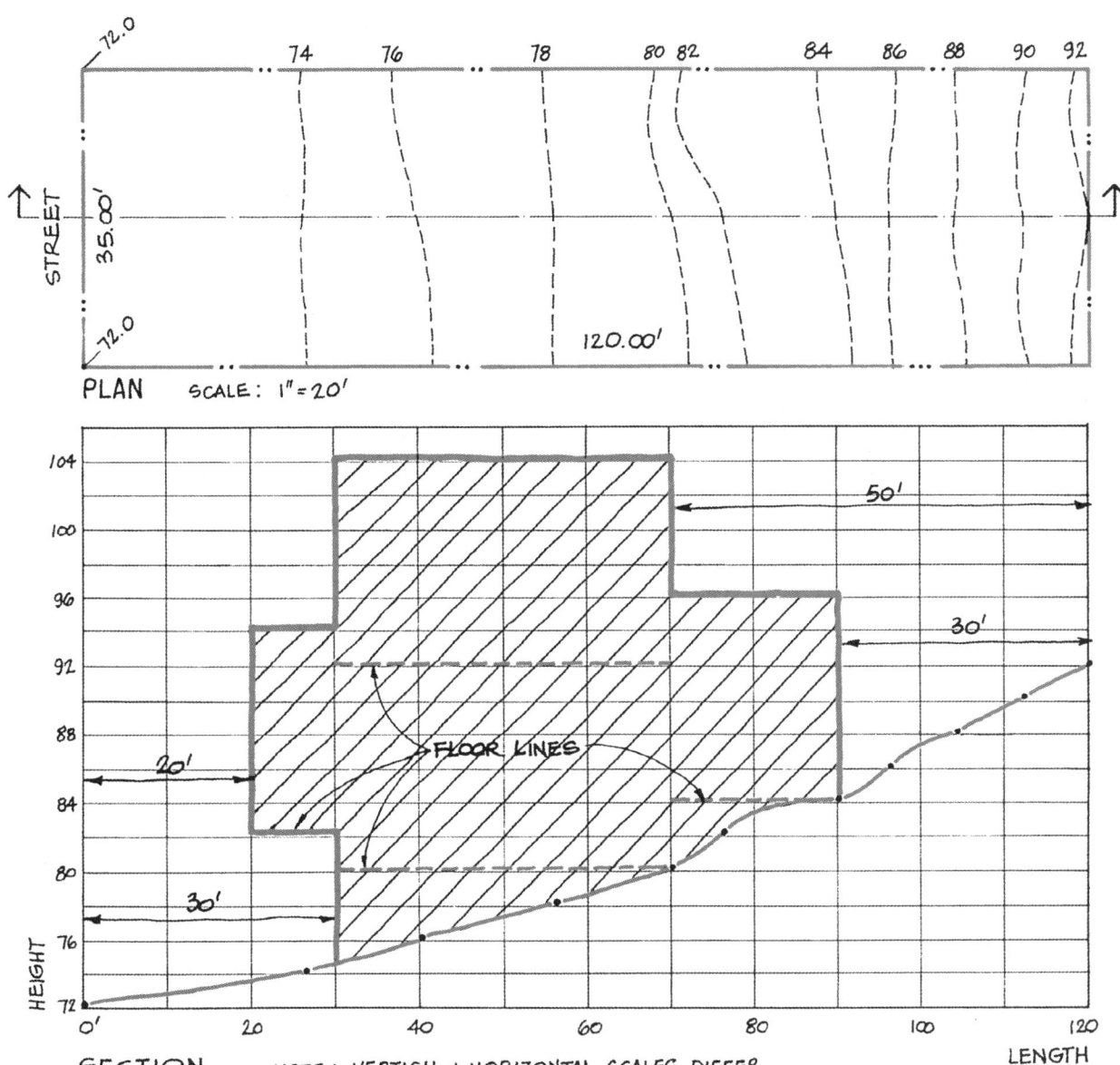

The next step is to determine the placement of 12-foot-high stories that fall within our established limit lines. We begin at the left, 20 feet distant from the street, where our lowest floor is already noted as 82. (Remember? 10 feet above the street level). From that point we mark off the 12-foot floor height along elevation 94. We may not mark off a second 12-foot story at that point, because our upper limit line at elevation 104 is only 10 feet above elevation 94, not the 12 feet that is required for each floor.

At what point, therefore, may we establish two 12-foot floors, one over the other? To determine that point we begin at our upper limit line at elevation 104 and measure downward 24 feet (2 × 12) to elevation 80. Thus, our lowest floor would be at elevation 80, and our second floor would be at elevation 92. However, this two-story condition may only occur between vertical lines 30 and 70. The 30 line represents the required setback previously established, and the 70 line is where floor elevation 80 meets natural grade. Remember, we are prohibited from constructing any part of the structure below existing grade.

Moving now to the rear of the property, we see the established 90 vertical line representing our required 30-foot setback. This line intersects the natural grade at elevation 84. We project up from that point 12 feet, to elevation 96, in order to establish the top of that rearmost volume. Again, we may not add a second story here, because there are only 8 feet remaining between the top of this story and our upper limit line at elevation 104.

We have now determined the location and extent of every 12-foot-high story that falls within the limit lines established earlier on our graph/section. The only remaining tasks are to outline and hatch the resulting buildable volume and provide all the front and rear setback dimensions at every floor level. This is accomplished as shown on our solution, and the problem is complete.

As one can see, the actual solution to this problem is not nearly as difficult as interpreting the written program. That is why candidates must proceed with such problems calmly, carefully, and always one step at a time.

PROBLEM 14 AIRPORT VAN TERMINAL

On the level site shown, a new transportation terminal will be built for airport vans. The facility will include parking, van loading, and passenger drop-off and pick-up area. Following are the requirements of this problem:

1. Terminal building, whose footprint is indicated.
2. On-site parking for 12 cars, including 10 - 10' × 20' spaces plus 2 - 15' × 20' spaces for the handicapped.
3. 10-foot-wide drop-off and pick-up platform along a 20-foot-wide driveway at the terminal entrance. Provide waiting space for a minimum of two cars.
4. 8 van spaces - 18' × 25' each - as indicated on the next page. None of these spaces may be farther than 40 feet from any part of the terminal.
5. 10-foot-wide loading platform between the terminal and the vans.
6. Minimum car dimensions:
    - Driveways - 20'
    - Outside turning radius - 35'
    - Inside turning radius - 15'
    - Back-up space - 25'
7. Minimum van dimensions:
    - Driveways - 25'
    - Outside turning radius - 40'
    - Inside turning radius - 20'
    - Back-up space - 40'
8. Passengers shall not cross van lanes to load or unload.
9. Site restrictions:
    - No more than four curb cuts are permitted, which must be a minimum of 50 feet from property lines at streets and 20 feet from side and rear property lines.
    - Setbacks are 20 feet at streets and 10 feet at side and rear yards.
    - No development may occur in the setbacks, except access driveways.
    - A 20-foot-wide landscaped buffer zone must be maintained at the north.
    - Do not disturb existing site vegetation.

Assignment:

Prepare a site plan showing the terminal location, all parking and van areas, and drop-off and loading areas. Dimension all vehicular areas, setbacks, and turning radii. Number all vehicle spaces and indicate traffic flow with arrows. Note that all auto traffic shall be one-way. Hatch all pedestrian traffic areas, and indicate some landscaping material.

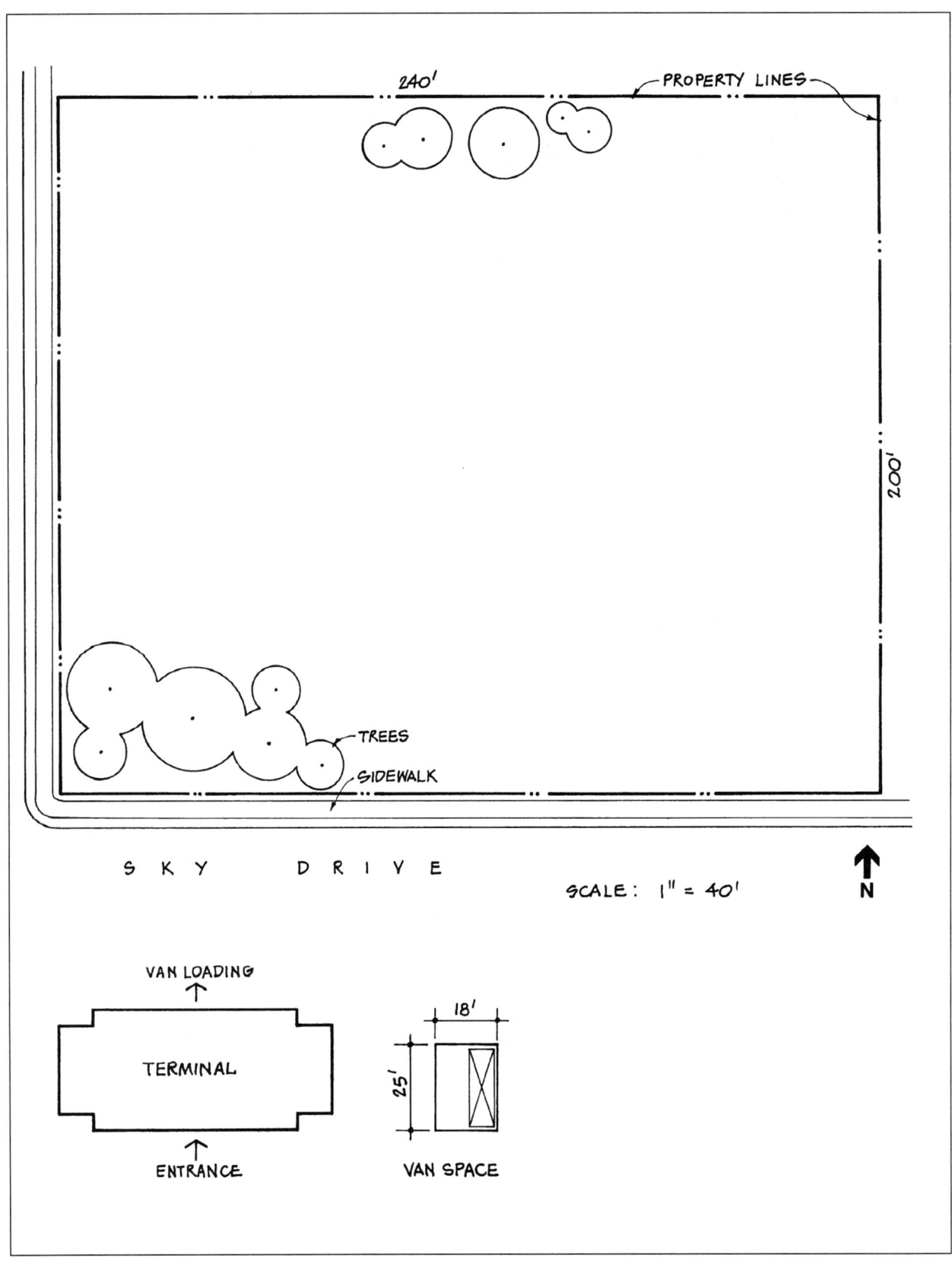

Problem 14: Discussion of Solution

The Site Planning exam consists of several vignettes that test technical skills and a more complex vignette that tests design logic. This problem is an example of the latter type, and although it requires a certain amount of technical skill to solve, it is really a complicated design problem. The exam is scored in such a way that a candidate who fails to solve the longer problem will very likely fail the entire test. Therefore, design problems such as this one are especially important to understand.

At first glance, this appears to be just one more parking problem, but on further review, candidates will discover that this is an intricate circulation problem that requires a number of critical design decisions. Further complicating this problem is a limited site incorporating numerous restrictions and a great many vehicles, together with their dimensions, all of which promises to make this a troublesome exercise.

Before beginning to sketch, candidates should try to understand what this problem is all about. Airport van terminals are similar to bus stations, in that they provide passengers with public transportation to and from an airport. Such terminals are centrally located, away from the confusion of an airport, so that drop-off and pick-up are uncrowded, efficient, and convenient. The sequence of events for a departing passenger would be as follows:

1. Enter the property and park or be dropped off at the terminal entrance.
2. Enter the terminal and purchase a ticket.
3. Exit the terminal to a loading platform, board a van, and depart for the airport.

The sequence of events for arriving passengers would be similar, except reversed.

Essentially, there are three major elements in this problem: The parking area, the terminal area, and the van area. It is essential to place these in proper functional order, so that the solution accomplishes the primary aim of the program. For example, it is clear that the terminal must be placed between the parking and van areas, because the terminal footprint indicates the entrance on one side (presumably where passengers are dropped off) and van loading on the opposite side. Thus, pedestrian circulation must be direct and continuous and resemble the following diagrammatic sketch:

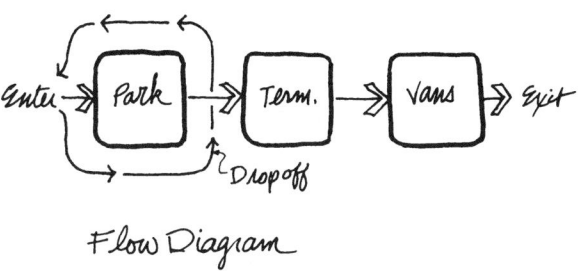

Flow Diagram

Simple though it may appear, this diagram is the critical key to a successful solution.

If we draw the setbacks on the site, we can see that the area remaining for development will measure 170' × 210' (10' + 20' = 30' subtracted from each site dimension). Since we have determined that our design solution will probably involve a linear layout, we decide to place the sequence of elements (as shown in our previous sketch) in the east-west direction, where we have the greater length. Therefore, the parking will be located adjacent to Air Street at the west.

We are told that auto traffic is one-way, which means that we will have two 20-foot-wide driveways, one entry and one exit. In addition, the placement of these driveways is restricted by the location of curb cuts, which must be a minimum of 50 feet from the south property line

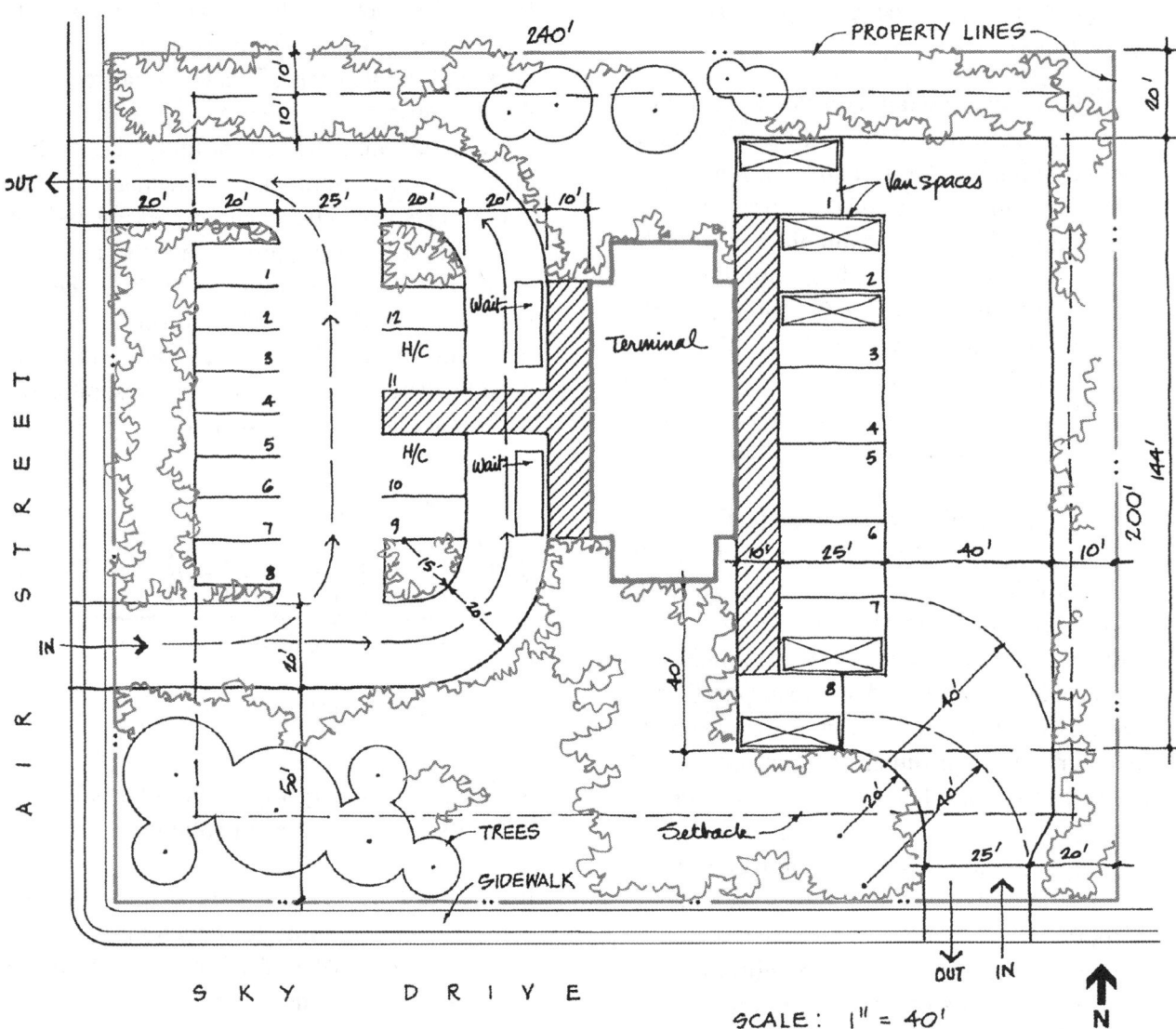

and 20 feet from the north property line. Therefore, our maximum development width along Air Street will be 200' − (50' + 20') = 130 feet. After deducting the width of two one-way driveways, we are left with 130' − (2 × 20') = 90 feet.

Assuming that end spaces require a few more feet of width than a typical parking space (for maneuvering), our dimension of 90 feet will yield a maximum of eight typical 10-foot-wide spaces.

It is apparent, therefore, that we will need two rows of parking, in order to accommodate the 12 required spaces. In our solution, we indicate a second row of cars, east of the first row, that includes the remaining two typical car spaces, and a 10-foot-wide pedestrian aisle at the center of the area, between the two required handicapped spaces. The two parking rows are divided by a 25-foot aisle (using the back-up, not the driveway, dimension).

The next element to deal with is the 20-foot-wide driveway adjacent to the 10-foot-wide drop-off platform at the terminal entrance. This circulation route is placed adjacent to the eastern parking row, and it is merely the continuation of (and link between) the two driveways that begin at Air Street. As the driveway circles around the parking area, we must be certain to comply with the turning radii specified in the program.

In reviewing the circulation pattern of this area, candidates should note that cars must enter the southerly driveway on Air Street, so that the passenger side of the car will face the terminal at the drop-off platform. Cars entering the site may go directly to the terminal, drop off their passengers, and then exit the northerly driveway, all in one continuous process. Or they may enter the site, park, and then proceed on foot to the terminal. However, because the circulation is one-way, they are unable to drop off passengers and then return to the parking area without first returning to Air Street. This is generally considered a disadvantage, but in an operation of this scale, the situation should produce no real hardship.

Next, the terminal building is centered on the east-west axis of the parking area so that the entrance is equidistant from the most remote cars in the lot (cars 1 and 8). The building is separated from the main driveway by the 10-foot-wide drop-off platform, which runs nearly the full building length. The pedestrian aisle in the parking area continues across the driveway (painted lines on the pavement) and terminates at the drop-off platform. So far, everything fits, it all functions properly, and we have only the van area left to plan.

Development east of the terminal begins with the required 10-foot-wide loading platform.

We were advised that passengers must not cross van lanes to load or unload. Therefore, it is clear that van lanes will have to be arranged perpendicular to the loading platform. Otherwise, we would end up with the following unacceptable situation:

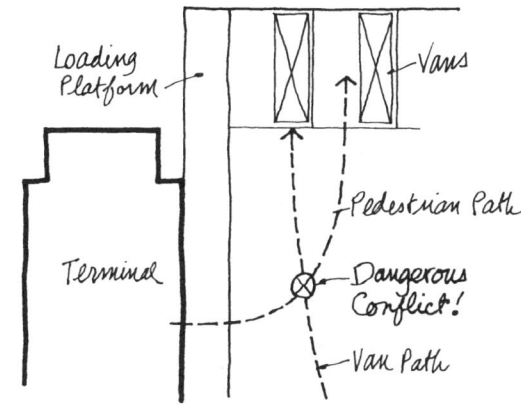

The most northerly van space is placed 20 feet from the northern property line (not the 10-foot setback), because we must respect the 20-foot-wide landscaped buffer zone at the north. From that point, we measure southward eight 18-foot-wide van spaces. We also measure eastward from the loading platform, first, 25 feet, which represents the van depth, and then the required 40-foot back-up space.

If you were to add all the setback and element dimensions, due east from Air Street to the rear of the van back-up space, the total would be 225 feet. This leaves 15 feet to the easterly property line (240' − 225' = 15'). This is the line along which the east side of the entry driveway should be placed in order to maintain the required 40-foot outside turning radius for vans. However, the program states that curb cuts are not permitted any closer to the eastern property line than 20 feet. Therefore, our 25-foot-wide driveway, from Sky Drive to the van area, must be offset 5 feet to the west as shown.

The resulting problem, however, is that now the outside turning radius for van space 8 can be only 35 feet, not the 40 feet required. This situation is illustrated below:

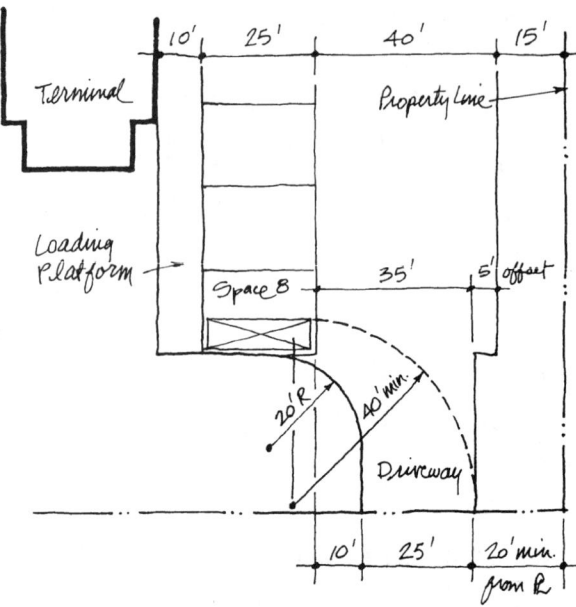

Our solution to this problem is to move van space 8 westward 10 feet, so that the loading platform terminates at the northern edge of space 8. A similar solution is used for van space 1, and this is done to effectively reduce the overall circulation distance.

There remains one more restriction to verify, and that is the requirement that no van space be farther than 40 feet from any part of the terminal. In fact, with the multitude of requirements and restrictions in this program, candidates should allow a few moments at the end to reread the program and verify every detail. This includes, for example, the provision for waiting space for two cars at the terminal entrance.

Our final concern is to note the required dimensions, vehicle space numbers, traffic arrows, and all other necessary notes. With a modest indication of landscaping, our solution is complete.

PROBLEM 15 AUTO CLUB BRANCH OFFICE

A new Auto Club branch office building is to be built on the site shown. A unique feature is the service window, where members can pay bills, obtain maps, etc.

Site:

- The lot slopes gently from southeast to northwest.
- Existing trees must not be disturbed.
- *Either* Olds Street *or* Chevy Avenue may be considered the front of the site.
- Setbacks are as follows:
  Front   40 feet
  Rear   20 feet
  Both sides   10 feet
- No development may occur closer than 30 feet from the adjacent Electric Department facility.

Parking:

- No parking is permitted within setbacks, easements, or 10 feet of the building.
- Standard parking stall is 10 feet × 20 feet.
- Handicapped parking stall is 15 feet × 20 feet.
- Aisles and driveways are 25 feet wide.
- A maximum of two curb cuts are permitted, and they must be located at least 100 feet from the intersection of the street centerlines.
- When backing out of parking spaces, no car may project into the principal circulation driveway.

Assignment:

Prepare a site plan indicating the following elements:
- Auto Club branch office, whose footprint is shown
- 10 public parking spaces *plus* one handicapped space
- 5 staff parking spaces
- Stack space for five cars at the service window
- 10-foot-wide bypass lane adjacent to stack space
- 10-foot-wide pedestrian walks from the building to the parking areas and to the street
- Retention pond, whose area is 10% of the site area and which may be located anywhere on the site
- Number all vehicle spaces, and indicate traffic flow with arrows

Site Planning

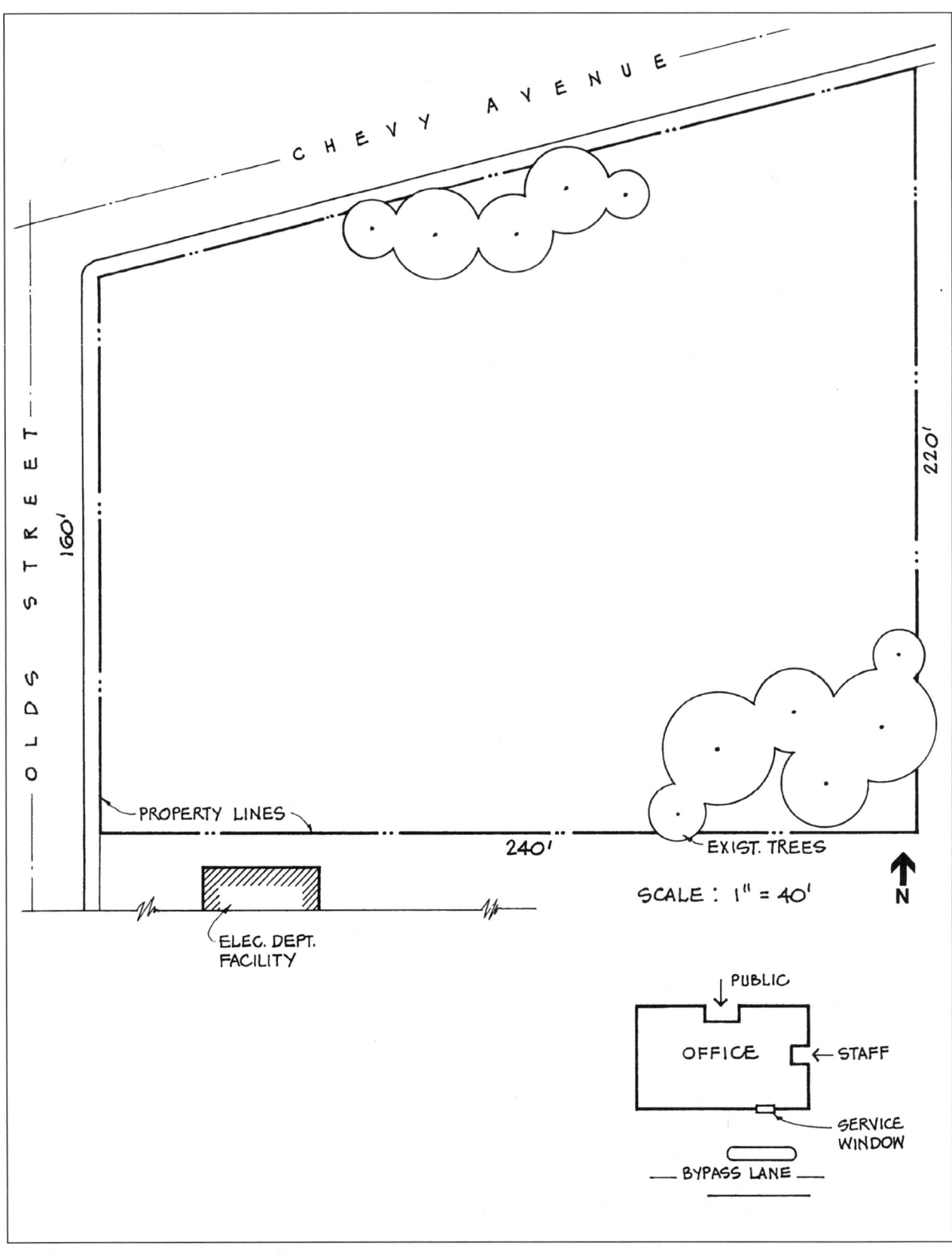

Problem 15: Discussion of Solution

This is another design logic problem that should take candidates at least an hour to solve. It is a complicated exercise involving parking, the circulation of cars and pedestrians, and a few important area calculations. One must read the lengthy problem carefully to avoid overlooking any requirement or restriction. For example, if you do not realize that traffic must approach the building with the driver's side of the car adjacent to the service window, your entire design will be in error. Even if you recognize such a mistake halfway through the exercise, it is unlikely that there will be sufficient time to redraw a correct solution.

After reading the problem and reviewing the site drawing, one may wonder how to begin such a complex circulation layout. Normally, one would start by laying out the setbacks, in order to determine how much space remains for development. But in this case, we don't even know which of the two streets is the front of the lot! That is a choice which must be made on the basis of how much area remains after the setbacks are deducted.

For example, if we assume Olds Street is the front, then the area is computed as follows: $[(160 + 220) \div 2 - (10 + 10)] \times [240 - (40 + 20)] = 170 \times 180 = 30,600$ square feet. If, on the other hand, Chevy Avenue is considered to be the front of the site, then the area that may be developed is computed as follows: $[(160 + 220) \div 2 - (40 + 20)] \times [240 - (10 + 10)] = 130 \times 220 = 28,600$ square feet.

Since the buildable area in the first case results in 2,000 square feet of additional area (ignoring the Electric Department facility easement), we consider Olds Street as the front of the site and lay out the required setbacks accordingly.

Next, we must consider the setback from the Electric Department facility at the south. The prescribed 30-foot dimension is measured both north and east from the facility (not the property line), in order to complete the outline of our site's buildable area.

A review of the problem statement indicates that vehicular circulation should be one-way, and because the driveway location is restricted (100 feet from the intersection of street centerlines), there will probably be one driveway located off each street. At this point, candidates should try to conceive the circulation pattern in general terms. For example, cars will enter the site and must be able to circulate to one of three destinations: the parking areas, the service window, or directly to the exit, in the event they wish to leave the site. This pattern is illustrated below.

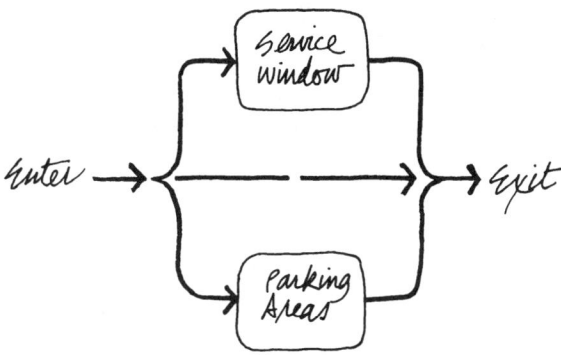

It is important to realize that cars at one destination must not interfere with the circulation of cars traveling to another destination. In fact, keeping all the traffic smoothly flowing and unrestrained is what this problem is about. You are being tested on your ability to arrange these various flow paths into an efficient and skillful solution.

The public parking area should be located as conveniently as possible to the office building. Therefore, the principal circulation driveway

224 Site Planning

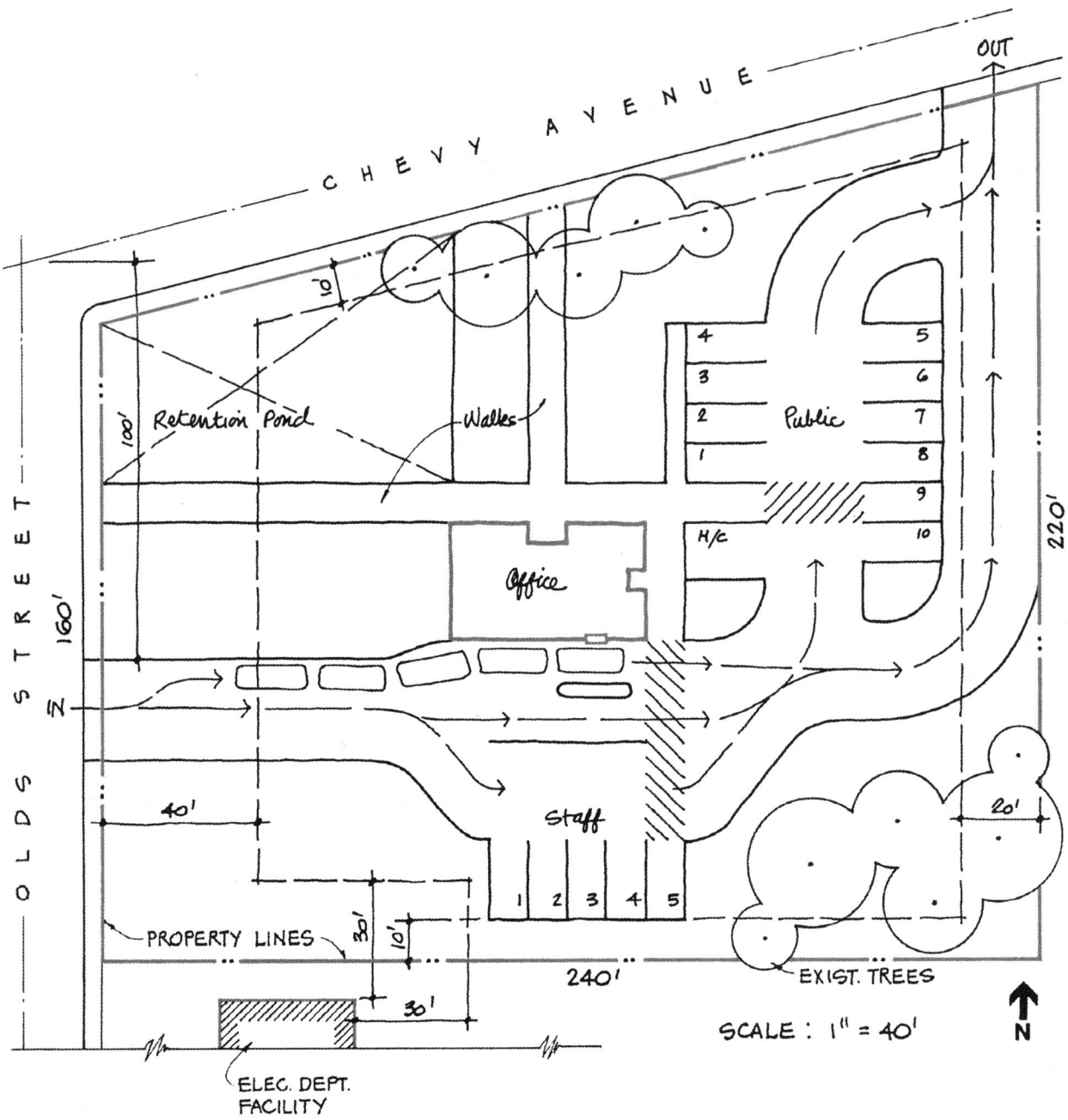

must run around the public parking area and be located further from the building than the parking. If we assume the entrance to the site is on Olds Street and the exit is on Chevy Avenue, our diagrammatic arrangement would be as follows:

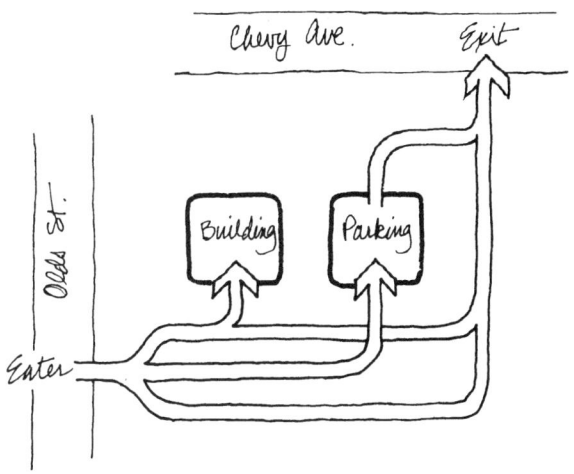

Although parking is prohibited within setbacks, no such restriction applies to driveways. Therefore, we begin our layout by placing the 25-foot-wide circulation driveway adjacent to the east property line. Following our diagrammatic plan and moving westward, we mark off a row of parked cars (20 feet), a parking aisle (25 feet), and another row of parked cars (20 feet). Since parking is not permitted closer to the building than 10 feet, we next mark off a 10-foot path and then locate the building with its long axis running east and west.

The layout of elements in the north-south direction begins with the staff parking area at the southerly setback line, in order to be separated from the public parking. Between the Electric Department facility easement at the west and the existing trees at the east, there is sufficient space to locate our five required staff car spaces. Moving northward, we mark off this row of cars (20 feet), a parking aisle (25 feet), and finally, the bypass lane, service lane, and office building indicated in the footprint.

The location of our entry driveway is dictated by the required 100-foot dimension from the intersection of the street centerlines. As such, the northerly edge of the entry drive does not align with the south face of the building, as we would prefer. However, a small jog in the driveway solves the problem without inhibiting a smooth flow of traffic. We indicate a car at the service window, and the remaining four cars are stacked behind at about 20-foot intervals. Finally, we add two 10-foot-wide pedestrian walks to the street, one centered on the building's public entrance, leading through the trees to Chevy Avenue, and the other adjacent to the north building face, terminating at Olds Street.

It should be understood that the layout shown here did not happen quite as easily as it has been described. Solutions of this kind require several trial-and-error layouts before all the elements fall into place. However, if one begins with a diagrammatic sketch that is logical and sound, he or she will ultimately end up with a plan that works. In that regard, candidates must be realistic about traffic patterns, circulation, and especially turning radii of cars. The turning radius used here is 15 feet for inside curves, which is an acceptable minimum.

There remains one more element to locate, and that is the retention pond, which is prescribed as 10 percent of the site area. To begin with, a retention pond is simply a large holding area for water which is used to detain runoff water on a property until the storm subsides. At that time, it may be disposed of through percolation into the earth or by slowly releasing the runoff to the public storm sewer system.

We are told that the lot slopes down towards the northwest, so obviously, the best location for our retention pond is at the lowest corner of the site. However, we must first compute the total lot area, which is done as follows: $[(160 + 220) \div 2] \times 240 = 45{,}600$ square feet. Thus, our retention pond area is $45{,}600 \times 0.10 = 4{,}560$ square feet. We begin at the northwest corner of the site and, through trial and error, describe a trapezoid about 90 feet long in the east-west direction and about 51 feet wide at the mid-point. Finally, we number all vehicle spaces, indicate the traffic flow with arrows, and add the appropriate notes that will help describe our solution.

Candidates may notice that in our solution traffic lanes may appear in conflict, such as when a car leaves the service window and attempts to exit the site. Obviously, with so many different circulation paths, some potential conflicts are unavoidable. Our advice is to minimize these situations and arrange the most reasonable solution within the restrictions of site and program. One must also remember that all design involves compromise, not only on the Site Planning test, but in the real world as well.

GLOSSARY

The following glossary defines a number of terms, many of which have appeared on past exams. While this list is by no means complete, it comprises much of the terminology with which candidates should be familiar. You are therefore encouraged to review these definitions as part of your preparation for the exam.

A

Access Right Right of an owner to have ingress and egress to and from a property.

Accessible Parking See Handicapped Parking.

Accessory Building A building or structure on the same lot as the main or principal building.

Aesthetics The study or theory of beauty.

Air Rights The rights to the use or control of space above a property.

Altitude The angle that the sun makes with the horizon.

Aquifer An underground permeable material through which water flows.

Azimuth A horizontal angle measured clockwise from north or south.

B

Barrier-Free The absence of environmental barriers, permitting free access and circulation by the handicapped.

Bearing In surveying, a direction stated in degrees, minutes, and seconds as an angular deviation east or west from due north or south.

Bearing Capacity The ability of a soil to support load.

Bench Mark A relatively permanent point of known location and elevation.

Berm A convex-shaped bank of earth.

Boundary The legal recorded property line between two parcels of land.

Buffer Zone An area separating two different elements or functions.

Buildable Area The net ground area of a lot that can be covered by a building after required setbacks and other zoning limitations have been accounted for.

Building Line A defined limit within a property line beyond which a structure may not protrude.

Building Envelope The enclosure that contains a building's maximum volume.

C

Catch Basin A drainage device used to collect water, with a deep pit to catch sediment.

Circulation The flow or movement of people, goods, vehicles, etc., from place to place.

Climate The generally prevailing weather conditions of a region throughout the year, averaged over a series of years.

Coefficient of Runoff A fixed ratio of total rainfall that runs off a surface.

Collector Street A street into which minor streets empty and which leads to a major arterial.

Combined Sewer Sewer that carries both storm water and sanitary or industrial wastes.

Comfort Zone Any combination of temperature and humidity in which the average person feels comfortable.

Compaction The reduction of soil volume by pressure from grading machinery.

Condemnation Taking private property for public use, with compensation to the owner, under the right of eminent domain.

Conduction The transfer of heat by direct molecular action.

Conduit Pipe or other channel, below or above ground, for conveying pipelines, cables, or other utilities.

Conforming Use   Lawful use of a building or lot that complies with the provisions of the applicable zoning ordinance.

Coniferous   Describing a cone-bearing tree or shrub. See Evergreen.

Context   The circumstances or elements which surround a particular development.

Contour   A line on a plan that connects all points of equal elevation.

Contour Interval   The vertical distance between adjacent contour lines.

Convection   The transfer of heat by the movement of a liquid or gas, such as air.

Corner Lot   A land parcel that fronts on two contiguous streets. The short side is generally considered to be the front of the lot.

Covenant   A restriction of the deed which regulates land use, aesthetic qualities, etc., of an area.

Cross Section   See Section.

Crown   The central area of a convex surface, such as a road.

Cul-De-Sac   A short road with an outlet on one end and a turnaround on the other.

Culvert   A length of pipe under a road or other barrier used to convey water.

Curb   A raised margin running along the edge of a street pavement, usually of concrete.

Curb Cut   A depression in a curb that provides vehicular access from a street to a driveway.

Cut and Fill   In grading, earth that is removed (cut) or added (fill).

D

Dead-End Parking   A circulation layout in which cars are unable to circulate in a continuous one-way flow from the entrance to the exit of a parking area.

Deciduous   Describing trees that shed their leaves annually, as opposed to evergreen.

Dedication   Appropriation of private property for public use together with acceptance for such use by a public agency.

Deed   A written instrument that is used to transfer real property from one party to another.

Degree Days   The number of degrees that the mean temperature for any day at a particular location is below 65°F.

Density   A measure of the number of people, families, etc., that occupy a specified area.

Discharge   Flow from a culvert, sewer, channel, etc.

Disposal Field   A system of trenches with gravel and loose pipes through which septic-tank effluent may seep into the surrounding soil. Also called Drainage Field or Absorption Field.

District   Any section of a city in which the zoning regulations are uniform.

Drainage   (1) The capacity of a soil to receive and transmit water. (2) The system by which excess water is collected, conducted, and dispersed.

Drainage Field   See Disposal Field.

Drip Line   An imaginary line on the ground described by the outermost branches of a tree.

Driveway   A vehicular path generally leading from a public street to a structure on private property.

Drop-Off   An area adjacent to a vehicular drive where pedestrians may safely exit (or enter) a car.

Dwelling Unit   An independent living area which includes its own private cooking and bathing facilities.

E

Earthwork See Grading.

Easement A limited right, whether temporary or permanent, to use the property of another in a certain way. This may include the right of access to water, light and air, right-of-way, etc.

Ecology The study of the pattern of relations between organisms and their environment.

Effective Temperature The sensation produced by the combined effects of temperature, relative humidity, and air movement.

Effluent Partially treated liquid sewage flowing from any part of a disposal system to a place of final disposition.

Elevation The vertical distance above sea level or other known point of reference.

Eminent Domain The right of a government, under the police power concept, to take private property for public use.

Encroachment Part of a building or an obstruction that extends into the property of another.

Envelope See Building Envelope.

Environment The natural and man-made things, conditions, and influences surrounding a person, community, or place.

Erosion The process by which the surface of the earth is worn away by the action of natural elements, such as water and wind. Also known as Weathering.

Evergreen Having green leaves throughout the year, as opposed to deciduous.

Excavation The digging or removal of earth.

Expansive Soil Clay that swells when wet and shrinks when dried.

F

Finish Floor Level The completed floor surface on which building occupants walk.

Finish Grade The elevation of the ground surface after completion of all work.

Flood Plain The land surrounding a flowing stream over which water spreads when a flood occurs.

Floor Area Ratio (FAR) The ratio of the floor area of a building to the area of the lot.

Flow Line The path down which water flows.

Front Yard The minimum legal distance between the front property line and a structure.

Frontage The length of a lot line along a street or other public way.

Frost Line The deepest penetration of frost below grade.

Function The natural or proper purpose for which something is designed or exists.

G

Geology The science that deals with the physical history of the earth.

Grade The elevation at any point. See also Gradient and Grading.

Gradient The rate of slope between two points on a surface, determined by dividing their difference in elevation by their distance apart.

Grading The modification of earth to create landforms.

Greenbelt A belt-like area around a city, reserved by ordinance for parkland, farms, open space, etc.

Greenhouse Effect The direct gain of solar heat, generally through south-facing glass walls and roofs.

Groundwater Level The plane below which the soil is saturated with water. Also called Groundwater Table or Water Table.

H

Hachure A shading technique used to depict ground form.

Handicapped Individuals with physical impairments that result in functional limitations.

Handicapped Parking Spaces designated for physically handicapped persons, consisting of a typical space with adjacent access aisle no less than five feet wide. Also known as Accessible Parking.

Humidity The amount or degree of moisture in the air.

Hydrologic Cycle See Water Cycle.

I

Indigenous Native to a particular region.

Infiltration The process by which water soaks into the ground. Also called Percolation.

Insolation The amount of solar radiation on a given plane.

Interchange The junction of a freeway with entering or exiting traffic.

Interpolation Determining an unknown value between known values.

Intersection The point at which two streets come together or cross.

Invert Elevation The elevation of the bottom (flow line) of a pipe.

L

Land Coverage The ratio of the area covered by buildings to the total lot area, expressed as a percentage.

Landscaping The conscious rearrangement of natural outdoor elements for function and pleasure.

Latitude The number of degrees north or south of the equator of a particular point on the earth's surface.

Legal Description Designation of boundaries of real estate in accordance with one of the systems prescribed by law.

Limit Line Any line beyond which development is prohibited.

Loop Street A minor street that comes off a major street, runs for a short distance, and then returns to the major street.

Lot Line The boundary line of a lot.

Lot Area Total horizontal area within the lot lines of a parcel of land.

M

Macroclimate The general climate of a region.

Manhole An access hole in a drainage system to allow inspection, cleaning, and repair.

Metes and Bounds A formal description of the boundary lines of a parcel of real property in terms of the length and direction of those lines.

Microclimate The climatic characteristics unique to a small area, caused by local features.

Multiple Dwelling A building containing three or more dwelling units.

N

Neighborhood A community of people living in a general vicinity. The area can generally support an elementary school.

Network A system of circulation channels which covers a large area.

90-Degree Parking A pattern of vehicle storage in which car stalls are arranged at a right angle to the access lane. Also known as Perpendicular Parking.

Non-Conforming Use A particular use of land or a structure which is in violation of the applicable zoning code. Generally, if the use was established prior to the code rule which it contravenes, it may continue to exist.

O

Off-Street Parking Space provided for vehicular parking outside the dedicated street right-of-way.

One-Way Traffic A circulation system in which all vehicles move in the same direction.

Open Drainage The removal of unwanted water by means of surface devices.

Orientation A position with respect to the points of the compass.

P

Pad An approximately level building area.

Parallel Parking A pattern of vehicle storage in which car stalls are arranged parallel to the access lane, as in conventional street parking.

Parking Lot An open space for the storage of motor vehicles.

Parking Stall A space in a parking lot marked off for the storage of a single motor vehicle.

Party Wall A wall built on the dividing line between two adjoining parcels, in which each owner has an equal share of ownership.

Passive Solar System A heating or cooling system that collects and moves solar heat without using mechanical power.

Percolation See Infiltration.

Perpendicular Parking See 90-Degree Parking.

Plane Surface A topographic configuration created by straight, evenly spaced contours.

Planting Strip A landscaped strip of ground dividing a pedestrian walk from a street.

Police Power The legal power of a government to authorize actions which are in the best interest of the general public.

Precipitation Water that falls on the land as rain or snow.

Principal Building A building that houses the main use or activity occurring on a lot or parcel of ground.

Property Line A legal boundary of a land parcel.

PUD A planned unit development, similar to a cluster development but larger in scale including, in addition to housing, commercial and industrial developments.

R

Radiation The process by which heat or other energy is emitted by a body, transmitted through space, and absorbed by another body.

Rational Method A method for computing approximate storm water runoff.

Rear Yard The minimum legal distance between the rear property line and a structure.

Relative Humidity The ratio of the actual amount of moisture in the air to the maximum amount of moisture the air could hold at a given temperature.

Restrictions Limitations on the use of property defined by covenant in deeds, by private agreement, or by public legislative action.

Retaining Wall A wall constructed of timber, masonry, or concrete designed to resist the pressure of the earth mass with which it is in contact.

Retention Pond An area used to retain and hold runoff water during a storm. The water is held until it is able to drain naturally.

Ridge A narrow convex land configuration represented by contours pointing downhill.

Right-Of-Way A strip of land granted by deed or easement for a circulation path.

Runoff The surface flow of water from an area.

S

Section The representation of a structure as it would appear if cut through by an intersecting plane to show its internal configuration. Also known as a Cross Section.

Septic System A sewage treatment system consisting of a tank and filtering system.

Setback The minimum legal distance between a property line and a structure.

Sewer An underground pipe or drain used to carry off excess water and waste matter.

Sheeting A thin layer of water moving across a surface. Also called Sheet Flow.

Side Yard The minimum legal distance between side property lines and a structure.

Silt A fine-grained soil whose particles are 0.05 to 0.002 millimeters in diameter.

Site Planning The art or science of creating or arranging the external physical environment.

Slope The inclination of a surface expressed as a percentage or proportion.

Sludge Accumulated solids that settle out of the sewage, forming a semi-liquid mass on the bottom of a septic tank.

Soil A natural material, formed of decomposed and disintegrated parent rock, that supports plant life.

Soil Boring Log A graphic representation of the soils encountered in a test boring.

Solar Zoning An ordinance controlling the mass and shape of buildings, which permits the penetration of sunlight between buildings.

Split Lot A lot that comprises more than one zone.

Spot Elevation The exact elevation at a key point on the ground or on a structure.

Spot Zoning Zoning that differs from the pattern of the surrounding area.

Stall See Parking Stall.

Story The vertical portion of a building included between the surface of any floor and the surface of the floor next above.

Subsidence The sinking of land.

Summit The highest point of a land mass, represented by concentric contours.

Sun Chart A map of the sky showing the path of the sun, from sunrise to sunset, on the 21st day of each month.

Surcharge Earth which is above the top of a retaining wall level.

Surface Water Water that runs along the surface of the ground, as opposed to below ground.

Swale A graded flow path used in open drainage systems.

Switchback Road A road that doubles back on itself with a hairpin curve.

T

Topography The configuration of the earth's surface.

Topsoil The upper six to eight inches of soil, which contains humus.

Transpiration The process by which water vapor escapes into the atmosphere from plants.

Trench Drain A linear drainage device used to collect and conduct water.

U

Uniform Slope A topographic configuration created by evenly spaced contours.

Utility Easement A legal right-of-way enabling a utility company to run service lines over private property.

V

Valley A narrow concave land configuration represented by contours pointing uphill.

Variance The special permission granted to the owner of a parcel of land waiving certain specific restrictions when the enforcement of these would impose an unusual or unreasonable hardship on the owner.

Vegetation All the plants, shrubs, and trees of a particular place.

W

Water Cycle The general pattern of movement of the water on, under, and above the earth. Also called Hydrologic Cycle.

Water Table See Groundwater Level.

Way Street, alley, or other thoroughfare or easement permanently established for passage of persons or vehicles.

Weathering See Erosion.

Windbreak A structure or plant which, because of its form and location, reduces wind velocities.

Wind Shadow The area near the bottom of the leeward side of a hill, where the wind velocity decreases to almost zero.

Y

Yard Open, unoccupied space on all sides of a building, based on the required setbacks.

Z

Zone Area established by a governing body for specific use, such as residential, commercial, or industrial use.

Zone of Aeration The zone below the ground in which the spaces between soil grains contain both water and air.

Zone of Saturation The zone below the ground in which all of the spaces between soil grains are filled completely with water.

Zoning The legal means whereby land use is regulated and controlled for the general welfare.

Zoning Ordinance Exercise of police power by a government in regulating and controlling the character and use of property.

BIBLIOGRAPHY

The following list of books is provided for candidates who may wish to do further research or study in Site Planning. Most of the books listed below are available in college or technical bookstores, and all would make welcome additions to any architectural bookshelf. In addition to the course material and the volumes listed below, we advise candidates to review regularly the many professional journals, which are available at most architectural offices.

ANSI A117.1 Handicapped Standards
American National Standards Institute
New York, NY

Architectural Graphic Standards
Ramsey and Sleeper
John Wiley & Sons, Inc.
New York, NY

Basic Elements of Landscape Architectural Design
Booth, Norman K.
Elsevier Science Publishing Co.
New York, NY

Design with Climate
Olgyay, Victor
Princeton University Press
Princeton, NJ

Environmental Analysis
Marsh, William M.
McGraw-Hill Book Co.
New York, NY

Landscape Architecture
Simonds, John O.
McGraw-Hill Book Co.
New York, NY

Landscape Planning for Energy Conservation
Environmental Design Press
Reston, VA

Principles & Practices of Grading, Drainage, and Road Alignment
Untermann, Richard K.
Reston Publishing Co.
Reston, VA

Site Planning
Lynch, Kevin
M.I.T. Press
Cambridge, MA

Site Planning Standards
De Chiara, Joseph
McGraw-Hill Publishing Co.
New York, NY

Solar Dwelling Design Concepts
AIA Research Corporation
Washington, DC

The Passive Solar Energy Book
Mazria, Edward
Rodale Press
Emmaus, PA

The Urban Pattern
Gallion, Arthur B.
Van Nostrand Co.
New York, NY

Time Saver Standards for Site Planning
DeChiara/Koppelman
Van Nostrand Reinhold Co.
New York, NY

Urban Design
Spreiregen, Paul D.
McGraw-Hill Book Co.
New York, NY

The Architect's Studio Companion: Rules of Thumb for Preliminary Design
John Wiley & Sons, Inc.
New York, NY

Design on the Land: The Development of Landscape Architecture
Belknap Press
New York, NY

Grade Easy
Landscape Architecture Foundation

Simplified Site Engineering
John Wiley & Sons, Inc.
New York, NY

INDEX

A

Applying for the test, 8
Aquifer, 25
Area drain, 45

B

Backup space, 190, 219
Bearing capacity, 25
Bench mark, 104
Berm, 193, 197
Bonus zoning, 81, 84
Bubble diagram, 96
Buildable area, 152, 181, 223
Buildable Area vignette, 127–130
Building and topography, 52–53
Building envelope, 109, 110, 178
Building placement, 161

C

Catch basin, 45, 195
Central inlet, 44
Circulation, 91, 190, 217, 223
Circulation systems, 60–75
Climate, 54
Conditional use, 81
Contours, 30–35, 102, 104, 110, 122, 145, 156, 166, 185, 186, 195, 206, 208
Contract zoning, 82
Crowns, 166, 200, 206, 208
Culvert, 46, 167, 202
Cumulative zoning, 75
Cut and fill, 36–41, 48

D

Detention pond, 23
Diagrammatic arrangements, 97
Drainage, 41–46, 145, 166, 200
Drainage field, 171
Drainage, roads, 206
Driveway, cross section, 146
Driveways, 141, 145, 166, 190, 192, 200, 217, 219, 225

E

Easements, 129, 152
Exam description, 1–2
Exam materials, 4
Exam room conduct, 12
Exam strategies, 125
Examination advice, 15–16
Existing land use, 55

F

Flexible zoning, 81–85
Floating zone, 81
Flood plain, 24, 171
Flooding, 23–25
Floor area ratio, 79–80
Flow lines, 145, 206, 208

G

Grade profile, 108
Gradient, 46, 116, 145, 156, 186, 206
Gradient percentages, 47
Gradient ratios, 46
Gradient, roads, 206
Grading, 36–41, 114, 185, 206
Grading Criteria, 3
Gutter, 44

H

Hachures, 35
Height limitations, 78
Hydrology, 55

I

Incentive zoning, 81, 84
Infiltration, 23
Interpolation, 195
Inverts, 206

L

Land coverage, 79
Limit lines, 89. *See also* Setbacks.
Linear pattern, 71

M

Managing problems, 15
Models, 35

N

Natural hazards, 55–56

O

Off-street requirement, 80
Orientation, 162

P

Pad, 156, 185
Parking, 51, 65–70, 190, 200, 217, 223
Parking aisles, 140, 190, 225
Parking Layout vignette, 132–136
Parking, handicapped, 200
Parking, parallel, 191
Pedestrian circulation design, 70–75
Percolation, 225
Planned unit development, 81
Plants, 26–29
Precipitation, 23
Property lines, 129

R

Ramps, 192
Retention pond, 225
Rezoning, 82
Roads, 50–51
Runoff, 23

S

Scheduling the test, 11
Section cut lines, 104, 110
Sensory qualities, 55
Septic system, 171
Setbacks, 89, 129, 152, 171, 172, 173, 178, 217, 223, 225
Setbacks and yards, 77
Sewer lines, 129
Shading, 33
Sheet flow, 146
Site analysis, 53–56, 96
Site checklist, 19
Site design, 1
Site design process, 59–60
Site Design vignette, 87–92
Site Grading vignette, 113–120, 143–147
Site Parking vignette, 141
Site planning test, 2
Site Zoning vignette, 101–111
Sloping plane, 43
Soil, 54
Solar access plane, 109
Solar zoning, 176
Spot elevations, 33, 193, 197, 201
Spot zoning, 82
Storm drainage, 41–46

Streams, 21–22
Sub-surface drainage, 44–46
Summit, 156
Surface drainage, 43–44
Swales, 43, 116, 147, 148, 159, 168, 208, 210

T

Taking the exam, 11–16
Time schedule, 14
Topographic maps, 29
Topography, 54, 156
Transpiration, 23
Trees, 88
Trench drain, 46
Turning radii, 219, 225

U

Underground water, 25
Utility easements, 130, 152

V

Valleys, 208
Variable setbacks, 78
Variances, 81
Vegetation, 55
Vehicular circulation, 62–65, 97, 140, 190, 223

W

Warped plane, 44
Water cycle, 23
Water wells, 171
Waterfalls and fountains, 22
Wetlands easement, 129

Y

Yards. *See* Setbacks.

Z

Zoning, 75–81
Zoning envelope, 77